Geometry for Grades K-6

Readings from the
Arithmetic Teacher

edited by
Jane M. Hill

National Council of Teachers of Mathematics

Library of Congress Cataloging in Publication Data:

Geometry for grades K-6.

1. Geometry--Study and teaching (Elementary)
2. Geometry--Study and teaching (Primary) I. Hill,
Jane M. II. National Council of Teachers of Mathematics.
III. Arithmetic teacher.
QA461.G46 1987   372.7   87-1715
ISBN 0-87353-237-6

The publications of the National Council of Teachers of Mathematics present a variety of viewpoints.
The views expressed or implied in this publication, unless otherwise noted,
should not be interpreted as official positions of the Council.

Printed in the United States

Contents

4. Geometry in Unusual Ways

5. Spatial Abilities

Preface

In 1970 the National Council of Teachers of Mathematics published *Readings in Geometry from the Arithmetic Teacher,* a collection of articles from the Council's official journal for mathematics instruction in grades K–8. Interest in geometry has increased since 1970, and the subject has become an important part of the curriculum at all grade levels, from kindergarten up.

The articles in this collection of readings come principally from the last ten years of the *Arithmetic Teacher*. Several criteria were applied in making the choices. First, the emphasis is on teaching mathematics in grades K–6. Individual articles may include teaching ideas and activities for grades seven and eight, but the focus is on the lower grades.

Because of the emphasis on primary and intermediate grades, articles on measurement geometry (finding the area and perimeter of plane figures and the volume of solid figures) are not included. Some of the activities lay the foundation for studying the formulas for area, perimeter, and volume, but the focus is always on the development of concepts of area, perimeter, and volume as characteristics of geometric figures. The emphasis is on the figures themselves rather than measurement.

The articles also feature a hands-on approach to the study of geometry. Informal geometry lends itself to *doing*. Children can enjoy many experiences with tangible materials: classifying geometric shapes in various ways, putting shapes together to create new shapes and patterns, building three-dimensional shapes with two-dimensional shapes, and so on. The "Let's Do It" articles, a regular department of the *Arithmetic Teacher* for a number of years (designated in the Contents by "LDI"), are examples of this approach.

Several articles describe activities that are appropriate for a sequence of grades. Ideas intended for one grade can often be adapted to a higher or lower one. Teachers can make their selections, depending on the grade and type of class they are teaching. One activity might be used just as the author describes it; another might suggest a technique that can be adapted to a different situation. Some articles may even enrich the mathematical knowledge of individual teachers. All articles were selected with the hope that the teaching techniques they describe will be useful to classroom teachers and beneficial to their students.

Introduction

GEOMETRY has a rightful place in the elementary school mathematics curriculum. The geometry appropriate to the lower grades, however, is informal, even though the subject itself is a very structured one. High school geometry is formally presented and studied. The facts surrounding points, lines, planes, triangles, cubes, cones, and so on, are organized and learned in a logical, developmental sequence. In fact, geometry's innate logic is part of its beauty.

The content of elementary school geometry is informal in organization, and it is presented in a way that allows students to *discover*. Classroom activities encourage children to observe, comment on, and compare their observations; make guesses about what they have observed; and then prove or disprove their guesses.

Why should geometry be an integral part of the elementary school mathematics curriculum? There are numerous reasons. Informal geometry—

- helps to develop children's spatial perceptions;
- sharpens children's observation and classification skills;
- fosters inquiry and discovery, and allows for individual differences in learning styles and interests;
- encourages children to make and test hypotheses;
- teaches basic geometric shapes and their distinctive characteristics;
- provides experiences with copying, creating, and extending patterns;
- lays the foundation for more formal geometry in the higher grades:
  - measurement geometry—formulas for area, perimeter, and volume; and linear, square, cubic, and angular measurements;
  - transformational and projective geometry and topology;
  - traditional Euclidean geometry;
  - fundamental geometric concepts—points, lines, line segments, rays, planes, space, and so on.

The articles that follow are loosely grouped into five categories: (1) geometric shapes, (2) blocks, (3) patterns and transformations, (4) geometry in unusual ways, and (5) spatial abilities. The articles do not cover every topic in the elementary school mathematics curriculum, but they do provide a sample of useful teaching techniques. They also cover topics that are unfamiliar to many classroom teachers.

Geometric Shapes

THIS section suggests activities using inexpensive and readily available materials that will involve children in thinking about triangles and circles in new ways. In "The Surprising Circle!" Dana and Lindquist present teaching ideas by grades 1–8. Teachers can choose ideas from several levels.

Damarin ("What Makes a Triangle?") describes activities with a miscellaneous collection of triangular shapes and paper strips cut from construction paper. In "A Triangle Treasury," Van de Walle and Thompson build activities around a set of fifteen triangular shapes. Detailed directions are included for making patterns for the figures, which can then be traced on construction paper and cut out. Dana and Lindquist suggest using triangle paper as a problem-solving approach to more experiences with triangles ("Let's Try Triangles").

Most of these activities can be adapted to other geometric shapes. Puzzles with circles and triangles can be adapted to puzzles with squares (tangram pieces, for example), rectangles, and even trapezoids. And what can be done with triangle paper can also be done with other types of graph paper.

These four articles also provide excellent examples of the discovery approach to teaching and learning mathematical concepts. The children are *doing* the mathematics.

Let's Do It

THE SURPRISING CIRCLE!

By **Marcia Dana**
Madison, Wisconsin
and **Mary Montgomery Lindquist**
National College of Education
Evanston, Illinois

What do you know about a circle? Often we consider only its measurements and fail to look at its other properties. What happens when you put several circles together or look at parts of a circle? Are looks deceiving? A circle looks so perfect and round, but can it fool you? Can it lead you on extensive searches? Can it produce unusual designs? What can you do with circular shapes and parts of circular shapes?

In the paragraphs that follow, we give clues to the answers to these questions by including one teaching idea for each grade level, from grade one through grade eight. Begin with the suggestion for your grade level and then look at the others. Can you encourage your fellow teachers to try going with circles? Could your school have a circle week, each grade level trying ideas and then sharing the results on the final day?

What ideas can you and your students come up with to add to the knowledge of the surprising circle?

Circle puzzles (Grade One)

Do you believe that the following pieces can make a circular shape when they are put together in a certain way?

Well, they can, as you will see. Puzzles shaped like circles are an easy and enjoyable way for first graders to begin exploring circles and circular shapes. You can quickly make many such puzzles, each different from the other.

1. Trace a large circle on a piece of construction paper. Cut out the circular shape. Use it as a pattern to trace as many other circles as you want, then cut out the circular shapes. Use a different color of construction paper for each circular shape so the puzzles will not get mixed up.

2. Cut each circular shape into three to six parts. For examples, see figure 1.

3. Each puzzle can be stored in an envelope or box. The puzzles can then be put out for the children to try whenever they wish. If these puzzles are too easy for some children, you could de-

Fig. 1

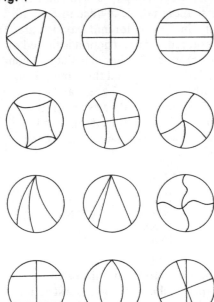

vise harder puzzles with more parts.

Another way to try circle puzzles is to have each child trace a large circle (using a can lid, perhaps) onto construction paper, cut out the circular shape, and then cut the circular shape into three pieces. Children could then trade puzzles with a neighbor and put together the neighbor's puzzle. The same thing could also be done with circular shapes cut into four or more pieces.

One other possibility, somewhat more challenging, is to trace a circle onto a rectangular shape, cut out the circular shape and discard it, and then cut the remainder of the rectangular shape into several pieces. Have the children put these pieces together, building around the missing circular shape.

You will probably think of other circle puzzles to try with your classes.

Circle Patterns (Grade Two)

Have you ever seen a strange pattern like the following?

Can you guess how it is made? As you will see, it and others can be made from fourths of circular shapes.

You will want to use several different colors of construction paper for the basic circular shapes so that color can be a part of the patterns. Trace many circles of the same size and cut out the circular shapes—the children can do this. Each circular shape must then be folded into fourths and cut apart on the fold lines. When you have a supply of fourths of circular shapes, all the same size but in different colors, the children can begin to put the fourths together, straight sides next to straight sides, to make patterns like those in figure 2.

Fig. 2

Notice that there are only two basic shapes for these patterns, although using color as part of the pattern makes them seem different. The children might want to make their patterns much longer than those shown in figure

Fig. 3

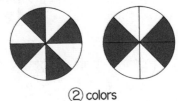

② colors

2. Also, some children will undoubtedly have other ideas about how to put fourths of circular shapes together to make patterns.

Another kind of circle pattern involves coloring a circular shape that has been divided into eight congruent parts. You will need to make a duplicating master of the basic figure.

You can do this by cutting out a circular region the size you want and then folding it into eighths. Use this as a pattern for tracing a circle and the fold lines onto the duplicating master. The children can then color in the sections in various ways.

Let the children make their own patterns. You might want to begin by suggesting that they use only two colors. Then they could try more colors. (See fig. 3.) This could also be tried with circular regions divided into six congruent parts, twelve congruent parts, and so on.

One other kind of circle pattern is made by tracing cans or lids in any way the child wishes. Circles of more than one size may be used (fig. 4). It is important to let the children use their imaginations in making these patterns, even if the result may not seem like a pattern to you.

Circular Rainbows (Grade Three)

What magic does it take to do the following?

 becomes

No magic, just *folding*. Through trial and error, children can find out how to

④ colors

Fig. 4

Using circles of the same size

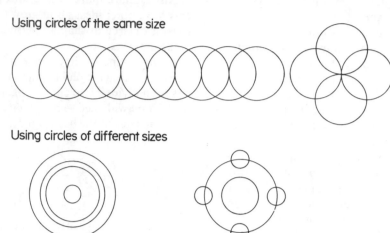

Using circles of different sizes

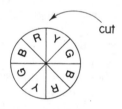

There are thirty-six ways to color circular shapes that are divided into eighths, using just one color and leaving parts uncolored. One restriction is placed on the coloring: One circular shape is considered to be colored the same way as another if the two shapes can be turned so that the colored parts are in the same position. For example, the following two ways of coloring are the same.

maneuver and fold the circular rainbows to make many different color combinations.

To begin, you will need some lightweight tagboard. Draw or trace a circle with a radius of about 10 centimeters onto the tagboard. Cut out the circular shape and fold it into eighths. Color the eighths as follows:

Color the other side the same way, so that each section is the same color on both sides—red over red, yellow over yellow, and so on. Now cut along a fold line, between red and yellow, to the center. (You can make as many circular rainbows as you wish, but you only need a few. The children need not do this in a large group.)

You will need to provide directions for the children, to show them what color combinations to try to fold and make with their circular rainbows. A large poster showing some or all of the combinations in figure 5 is one way of doing this. Before the children fold their circular rainbows to make the pictured color combinations, let them try, by folding only on the folds, to make any color combinations they wish. The children need to find out that there are lots of ways to fold the circular rainbows and that a fold line can be folded in either direction. Tell the children that their tagboard must be flat after

they have folded it into the color combination they want.

Once a child can make most of the color combinations shown on the poster, he or she can try some combinations using just the words as directions:

> green-blue-blue-green
> green-yellow-yellow
> red-blue-blue
> green-red-green
> blue-blue

The children can also make up color combinations and challenge one another to see if the color combinations that they suggest can be made. Some color combinations are not possible with this circular rainbow.

Circle Search (Grade Four)

Challenge your class to see if they can find all thirty-six circular patterns. What thirty-six circular patterns?

It may help to cut out the circular shapes to check. With this restriction and some persistent searching, your class should arrive at the thirty-six. Without the restriction, they will be searching a lot longer. You will want to make a duplicating master of circles divided into eighths (using radius of 2 cm you can get as many as sixteen on a sheet). Each child will need several copies of the figure.

To help the children in the search, guide them in the direction of being more systematic. Ask questions like the following:

How many ways can you color one part? Two parts? Have we found all of these?

What other numbers of parts can be colored? (Look at each of these separately.)

What about the positions of three colored parts? Are they next to each other? If two colored parts are next to each other, where could the third be? If

Fig. 5

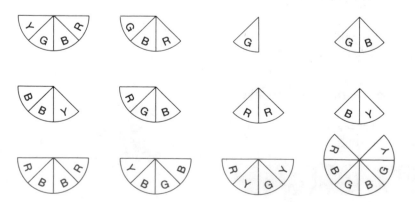

none of the three colored parts are next to each other, where could they be?

Your own enthusiasm and willingness to participate in the search will motivate the children, and the systematic search will give them a valuable experience in one technique for solving problems. If the children enjoy this type of search, you might try one with circular shapes divided into sixths (easier) or tenths (harder). Or you might find out how many ways a circular shape divided into halves or thirds can be colored with two colors.

Circle Designs (Grade Five)

Would you believe that from circles divided into eight congruent parts by dots all of the following can be made?

Five triangles of different shapes
Seven quadrilaterals of different shapes (including a rectangle and a square)
Five pentagons (five sides) of different shapes
Four hexagons (six sides) of different shapes
One heptagon (seven sides)
One octagon (eight sides)
And complex and not so complex designs such as the following:

To begin, each child needs to trace a circle with a radius of 3 to 4 centimeters. The child then cuts out the circular shape and folds it into eighths carefully, marking a dot where each fold line intersects the circle.

Then each child traces the original circle several more times and uses the marked figure to help mark dots on each newly traced circle. There is no cutting out of figures in this activity. Each child should keep the tracing circle and the folded, marked circle so as to be able to make more marked circles whenever he or she runs out.

Now you can ask each child to use a ruler to draw any two lines from a dot to a dot on one circle. For example, the following figures might be drawn.

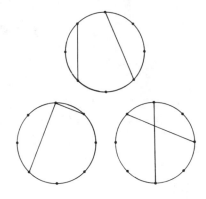

Have the group examine each child's two-line design to see how many different designs they have come up with. If the children are interested, they can look for more designs. There are at least thirty-four different possibilities (and this isn't counting designs that can be turned or flipped so they are the same).

Once the children have tried the two-line designs, challenge them to find the following figures:

A triangle with two congruent sides
All the different triangles that can be made in these circles
A square
A rectangle
A four-sided figure with three congruent sides

If the children are excited about the figures they have tried so far, you can challenge them to find all the other figures listed at the beginning of Circle Designs.

Then let the children use what they have learned so far to make their own designs, using as many lines as they wish. Their imaginations should take over at this point.

There is another way these same marked circles can be used. Instead of connecting the dots with straight lines, the children can connect them with curves. This can be done as follows:

1. Make a circle with eight equally spaced dots as before.

2. Take the circular shape that has been used for tracing circle and put it over the circle so that the edge of the tracing circle connects two dots on the circle. Trace the curve from dot to dot.

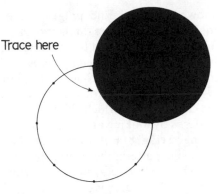

Trace here

There are three different curves that can be made in this way.

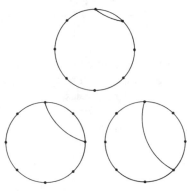

The curves can be used just as the straight lines were used to make various designs, like the ones in figure 6. Let the children discover different designs made by tracing the curve of the tracing circle. The children's experiences with straight-line designs should be a source of many ideas for them.

Fig. 6

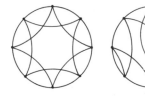

Circuzzles (Grade Six)

Have you ever met a circuzzle? Good, we hadn't either, until some circuzzles came around to a sixth grade, and soon the children were introducing us to more. A circuzzle is a puzzle made from fourths of circular shapes. Figure 7 shows some circuzzles that will start the children thinking about how fourths of circular shapes will fit together. Have each child draw two circles of the same size (radius about 6 cm), cut out the two circular shapes, and then fold and cut the circular shapes into fourths. The children are now ready to make the circuzzles shown in figure 7. Encourage them to make other circuzzles on their own. They might even want to try circuzzles made with more than eight pieces.

After children have had a chance to try these circuzzles, you can move on to making others in which children create patterns by looking at how the curve changes. (Fig. 8) For example, in the following pattern, the curve is put together for two pieces, and then the direction of the curve is changed.

Fig. 7

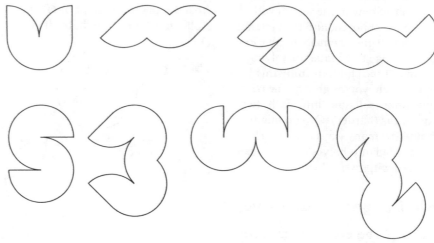

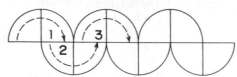

In this next example, the circuzzle will come back on itself. The curve is put together for three pieces, then the direction is changed for one piece, and then is changed again for three pieces together.

Fig. 8

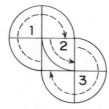 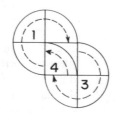

Have the children try some patterns of their own. They also can try using circular shapes cut into sixths or eighths. An interesting variation is to cut the circular shape into sixths as follows:

The children can also try to find how many different shapes they can make from two, three, four, or five fourths of circular shapes. (This is a variation of polyominoes.) The only rules are the following:

1. Two pieces are put together so that straight sides are together.
   Like this

Not like this

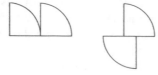

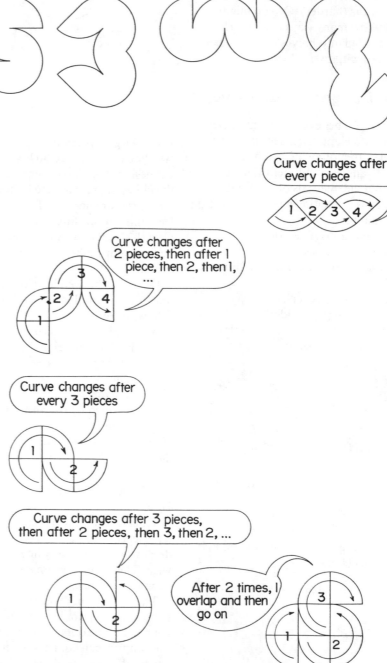

2. One shape is the same as another if it can be turned or flipped and then coincides with the other.

3. No piece can overlap another. If you have children do this, begin with two of the fourths and have the children find how many shapes can be made with two fourths. Then try three of the fourths. If children are having difficulty making different shapes, have them add one fourth, in different positions, to the shapes made by two

Fig. 9

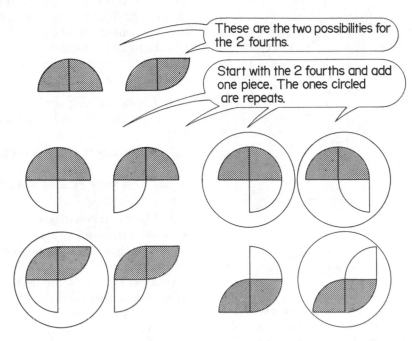

These are the two possibilities for the 2 fourths.

Start with the 2 fourths and add one piece. The ones circled are repeats.

fourths (fig. 9). For each different position of the third fourth, they should check to see if the shape they get is the same as a shape that has already been made. It may be necessary to tape the pieces together so the children can flip or turn the shapes. There are three different shapes that can be made from three fourths, six shapes that can be made from four fourths, and seven shapes that can be made from five fourths.

Circle Paper (Grade Seven)

Have you ever wondered what would happen if you made graph paper with circles instead of squares? (See fig. 10.) The best part is all the strange figures that can be made by tracing over some of the curves. Shapes like the following, as well as many others, can be made just by using the circle graph paper.

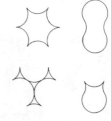

You will need many sheets of circle paper—students use it up fast. Make a sheet of circle graph paper, then make a duplicating master. (Students who do neat work with compass and ruler could make a sheet of circle paper. Notice that each circle is surrounded by six circles. The centers of the surrounding circles are on extended diameters of the circle in the middle.)

Figure 11 shows a variety of figures that can be made from circle paper.

Fig. 11

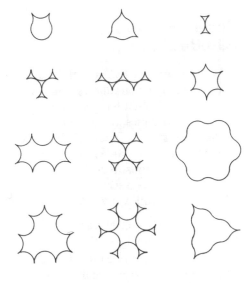

Most of the time, if they turn the circle paper face down, students will be able to see the circles well enough to trace over the curves on the blank side of the paper. Some students will want to make their own designs. They could also try to make circle numerals or circle letters.

Have you ever made wallpaper? You can have your students use their circle paper to try the designs in figure 12. Or they can make up their own designs. Coloring the wallpaper designs will make them look professional.

Fig. 10

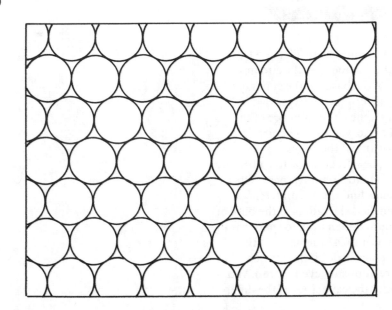

Fig. 12

  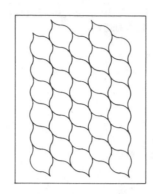

Circles to Cones (Grade Eight)

What happens when you make different cones from circular regions of the same size? What happens to the heights of the cones, the circumferences of their bases, to their capacities? Small groups of students can explore these and other questions. First have each small group make a set of cones according to the following directions:

1. Draw eight circles, each with a radius of 8 centimeters.

2. Divide each circle into eight congruent parts and draw the diameters connecting the equally spaced points on the circle. Do not fold on these lines.

3. Cut out each circular shape and cut along one radius of each.

4. Make a five-part cone by sliding three parts underneath the others and taping.
Then try a four-part cone.

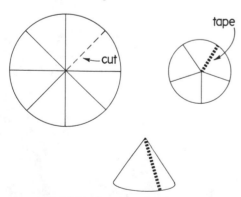

5. Now try all the other cones. When you are making the three-part cone, cut away four parts and slide the one extra part underneath, then tape. For the one-part cone, cut away six

parts; and for the two-part cone, cut away five parts. What happens with the eight-part cone?

Once the various cones are made, the small groups can look at their sets of cones to see what patterns can be observed. Then you may want to have the students investigate some of the properties of the cones. For example, how does the height change from the one-part cone to the seven-part cone? The height can be measured by holding the cone between two books.

Since the cones were all made from circles of the same size, the circumferences of the bases of the cones can be easily determined. (Expressed in terms of pi, the circumference of the original circle was 16π. The circumference of any of the cones would, therefore, be a fraction of 16π.) Is there a pattern in the circumferences? Once the students find the circumferences, have them find the radius of the base of each cone and then look for patterns in the radii. (Drawing the bases of the cones as concentric circles may emphasize the relationship even more.) And what about the capacities of the differ-

ent cones? If the cones are taped well, they can be filled with rice or sand and then the contents can be poured into a graduated liter container.

There are other variations of the circles-to-cones activity:

1. Divide the basic circles into six or ten congruent parts and proceed as before.

2. Investigate the five-part cones of eight-part circles of various sizes. Try circles with radii of 6cm, 7cm, 8cm, 9cm, and 10cm.

3. Make and investigate pyramids. To make a pyramid, proceed as if you were making a cone, dividing the circle into congruent parts and drawing the diameters. Then draw the chords connecting successive points on the circle.

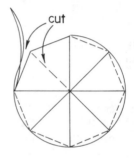 

Cut along the chords and fold along the diameters, and then continue as you would in making a cone. □

What Makes a Triangle?

By **Suzanne K. Damarin**

The triangle is one of the basic shapes to which children are exposed quite early in the curriculum, and often even before they enter school. However, although children learn the word "triangle" and some of the properties of triangles in kindergarten or first grade, their concepts of a triangle often remain very meager throughout the elementary grades.

Many children are familiar with only a few triangles (figs. 1a, 1b, and possibly 1c). Their concepts of triangles do not include triangles like those in figure 2, although they may know formally that a triangle has "three sides and three corners." Even when young children recognize a wide variety of triangles, they may not "see" the relationships among the lengths of the sides of a triangle. These relationships are summarized by the *triangle inequality,* an important geometric theorem which asserts that for any triangle, the length of any one side is less than the sum of the lengths of the other two sides.

The purpose of this article is to suggest several activities that can be used to give children more experience with triangles. With construction paper, a good paper cutter, and an ample supply of paste you can make materials for a variety of activities that can help broaden a child's conception of what a triangle is.

Suzanne Damarin is on the faculty of the Ohio State University where she teaches courses and conducts workshops for teachers on mathematics and the teaching of mathematics. She is involved in several mathematics curriculum projects including a United States Department of Education contract to develop curricula using microcomputers for elementary school mathematics.

Using Triangles in Designs

Using construction paper of various colors, cut a large supply of trianglar pieces in many shapes and sizes. Have children select some of these pieces and paste them on a large page to create a picture. Young children will enjoy making a "triangle man" or a "triangle cat" by working with unusually shaped triangles (fig. 3). Third graders may find this type of activity more interesting and challenging if you give them a region—bounded by a large rectangle or triangle, for example—to cover completely with non-overlapping triangular pieces (fig. 4). Discussion following construction of these figures should focus on identifying triangles in the pictures. In the process of finding triangles children's mind sets, such as the inclination to find only triangles with horizontal bases, will be broken.

The activity just described involves

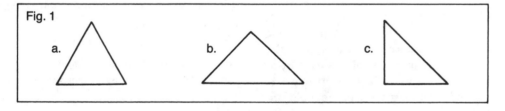

Fig. 1

a.   b.   c.

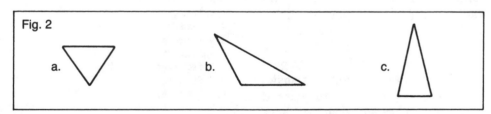

Fig. 2

a.   b.   c.

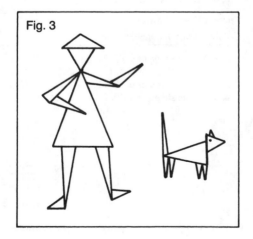

Fig. 3

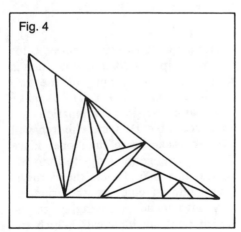

Fig. 4

children in manipulating diverse triangles. Problems such as "count the triangles" can be presented at all grade levels. In the early grades simple figures such as those in figure 5 might be used. In the upper grades more complex figures (fig. 6) can be used. Some teachers have had contests in which the person who has found, and can show the class, the largest number of triangles is the winner.

Making Triangles with Paper Strips

Cut paper strips one centimeter wide from seven different colors of laminated construction paper. (The lamination minimizes tearing.) Using a unit of 2 or 3 centimeters, make strips of length 3, 4, 5, 6, 8, 10, and 12 units. Each length should have its own color. For example, you might make all 3-centimeter strips red, all 4-centimeter strips blue, and so on. Near each end of each strip punch small holes. These strips can be used in a series of triangle construction activities that help children understand the relationships between the sides of a triangle and introduces them to similarity.

First activity

Give each child three strips of the same color to arrange to form a triangle. Make sure that each color is used by at least one child. When the triangle is formed, the corners can be held together with paper fasteners, wire staples, or paste. (If you are using staples or paste, be sure children form the full triangle before they fasten the strips.) Have the children compare their triangles. Make a nest of triangles (fig. 7) showing that these triangles fit inside each other and that they have the same shape.

Second activity

Give each child two strips of one color and one strip of a different color that is shorter that the length of the two matching strips together. Try to make sure that all possible combinations are used (see table 1); some children will need to make more than one. Have children arrange the strips to form triangles and then fasten the corners. Help children compare the triangles they have made by asking questions: Which ones are identical? Which tri-

angles have the same shape but different sizes? With which triangles can you make nests?

Third activity

Repeat the second activity, but this time have the children make triangles that have sides of three different lengths. (Table 1 can guide you in selecting sides of appropriate lengths.)

The class can make a bulletin board on triangles as a project. Assign each of the eighty-four combinations of three strips to a child. Each child should get two, three, or four combinations, depending on the size of your class. Make sure that each child gets one combination that does not form a triangle, even if you have to use some combinations twice. If the children can use rulers, ask them to measure the strips and mark the lengths on them. If the strips can form a triangle, the child should

construct it. If not, have children arrange the strips to show that two sides do not meet. Arrange the triangles in families or nests on the bulletin board. Help the class make observations about the combinations of lengths that do and do not yield triangles, about the relationship among triangles in a family, and so on. The results of these observations can be displayed on the bulletin board, too (see fig. 8).

Fourth activity

This activity can be done as brief "one-shot" experiences or expanded into larger projects.

Give each child some long strips of paper and three numbers such that the sum of any two is greater than the third. The numbers are to be lengths of the paper segments to be used in attempting to construct triangles. Different children can have different sets of

---

Table 1

Using strips with lengths 3, 4, 5, 6, 8, 10, and 12, there are 84 possible combinations of 3 strips. Triangles can be made from some, but not all, of these combinations. This table tells which combinations yield triangles and classifies the triangles according to shape.

Three identical strips

This combination will always give an equilateral triangle; all equilateral triangles have the same shape.

Two identical strips and one different

Combinations that yield isosceles triangles		Combinations that do not yield triangles
3, 3, 4 same shape as 6, 6, 8	5, 5, 8	3, 3, 6
3, 3, 5 same shape as 6, 6, 10	8, 8, 3	3, 3, 8
4, 4, 3 same shape as 8, 8, 6	8, 8, 4	3, 3, 10
4, 4, 5 same shape as 8, 8, 10	8, 8, 5	3, 3, 12
4, 4, 6 same shape as 8, 8, 12	10, 10, 3	4, 4, 8
5, 5, 3 same shape as 10, 10, 6	10, 10, 4	4, 4, 10
5, 5, 4 same shape as 10, 10, 8	10, 10, 5	4, 4, 12
5, 5, 6 same shape as 10, 10, 12	12, 12, 3	5, 5, 10
6, 6, 3 same shape as 12, 12, 6	12, 12, 4	5, 5, 12
6, 6, 4 same shape as 12, 12, 8	12, 12, 5	6, 6, 12
6, 6, 5 same shape as 12, 12, 10		

Three different strips

Combinations that yield triangles		Combinations that do not yield triangles	
3, 4, 5 same shape as 6, 8, 10	3, 6, 8	3, 4, 8	4, 5, 10
3, 4, 6 same shape as 6, 8, 12	3, 8, 10	3, 4, 10	4, 5, 12
3, 5, 6 same shape as 6, 10, 12	3, 10, 12	3, 4, 12	4, 6, 10
4, 5, 6 same shape as 8, 10, 12	4, 5, 8	3, 5, 8	4, 6, 12
	4, 6, 8	3, 5, 10	4, 8, 12
	4, 8, 10	3, 5, 12	5, 6, 12
	4, 10, 12	3, 6, 10	
	5, 6, 8	3, 6, 12	
	5, 6, 10	3, 8, 12	
	5, 8, 10		
	5, 8, 12		
	5, 10, 12		

---

numbers. Also give the children, or have them find, objects such as a paper clip, a piece of chalk, an eraser, a pencil, or a comb that can be used as a unit to measure lengths. Ask the children to make strips of the lengths given to them and then form triangles. For example, a child with lengths 2, 6, and 7 might form a triangle from strips, 2 paper clips, 6 paper clips, and 7 paper clips long (fig. 9). The same child could form another from strips, 2 combs, 6 combs, and 7 combs long. Help children to observe that all triangles made using the same numbers are the same shape and have them make nests of triangles. Have all children working with the same numbers compare and combine their nests. Compare the resulting nests with those on the bulletin board and with each other. (Be sure that this activity includes some number combinations not used in earlier activities—combinations involving 7 are good.)

Fifth activity

Ask the children to use paper strips of various lengths to invent a triangle of their own. Have the children compare their triangles: Which is the skinniest triangle? Which is the biggest triangle that is the same shape as Molly's triangle? The smallest? Which triangle has a square corner? Which triangle is a different shape from any other in the room?—You and the children will have other ideas as well.

Summary

As these activities illustrate, children can experience a diversity of triangles in a very concrete way—by handling triangular shapes and building triangles with strips. They can have first-hand experience with the idea that a triangle is completely determined once the lengths of its sides are fixed. They learn (some "the hard way") that you need to fit all three sides together at a random angle and to be sure the third side will fit. They also become aware that although every triangle has exactly three sides, you cannot always form a triangle from three strips. Examination of the triangles and nontriangles in the bulletin-board activity helps children develop an understanding of the meaning of inequalities such as $3 + 4 < 8$.

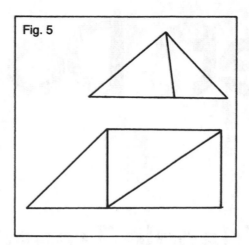

Fig. 5

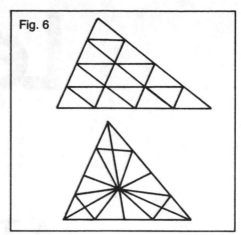

Fig. 6

Fig. 7

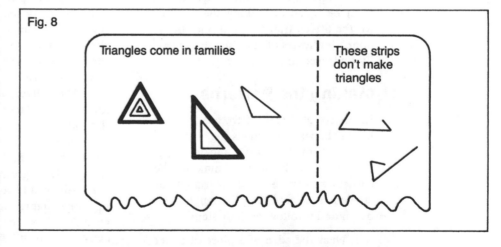

Fig. 8

Triangles come in families

These strips don't make triangles

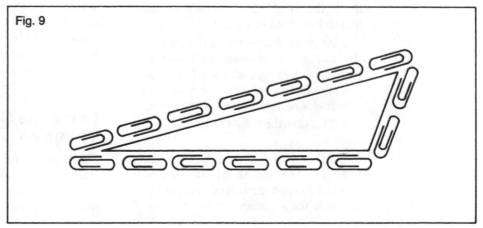

Fig. 9

Let's Do It

A Triangle Treasury

By **John Van de Walle,** *Virginia Commonwealth University, Richmond Virginia* and **Charles S. Thompson,** *University of Louisville, Louisville, Kentucky*

In this article, fifteen triangle shapes cut from poster board or construction paper provide the springboard for discussions of a variety of geometric concepts at all levels of sophistication, from grades 3 to 8, with many activities suitable for K to 2. Some time and care must be taken in making the patterns for the fifteen triangles, but once this is done, duplicates of each shape can easily be traced and cut.

Making the Patterns

In the set of fifteen, there are five different shapes of triangles (fig. 1) with a small (S), medium (M), and large (L) version of each shape. Think of the corners or vertices of each triangle as labeled *A*, *B*, and *C*. All triangles can be made by following four steps:

1. Near the edge of a piece of paper draw the side *AB* according to the length indicated in table 1.

2. At point *A* construct an angle with the number of degrees indicated in table 1 in the column labeled ∠A. The corresponding angles for the S, M, and L triangles of each shape are the same.

3. Measure the side *AC* according to table 1.

4. Connect *B* and *C* to complete the triangle. The length of $\overline{BC}$ can be checked against the length indicated in the table for accuracy.

Figure 2 illustrates these steps for making the triangle patterns.

From the paper patterns a poster-board set of triangle shapes can be cut. Simply trace the patterns onto poster board that is colored on both sides. Or, by using the patterns to make spirit masters, many sets of triangles can be made by feeding construction paper through a duplicating machine. All three sizes of each triangle shape will fit on one spirit master. Make the set of triangles in five colors with the small, medium, and large sizes of each shape the same color. Label the triangles S, M, or L. Each triangle can then be designated by size and color—medium red triangle, large green triangle, small yellow triangle, and so on. (Strictly speaking these figures are not triangles—a triangle is a set of line segments. The activities that follow focus on the characteristics of triangles. Thus, in the rest of the article we shall be referring to triangles without always distinguishing between the triangle shape and the triangle, unless that distinction is significant in the point that is being made.)

Using the Set of Triangles

The triangles are most useful when the class is separated into groups of from three to six students, with each group having a full set of the fifteen triangles.

Occasionally groups will want to share triangles so they can have two or three identical shapes. You may also wish to have extra sets available. All groups can work simultaneously under the direction of the teacher.

The triangles form the basis for explorations and open-ended questions directed by the teacher. Some discussions will lead to other drawings on paper or the chalkboard, measurements with rulers or protractors, cut-and-paste activities, and the use of grid paper and other materials.

Give some order to observations and discoveries. This may take the form of lists made on the board, experience stories written by the teacher, or notebooks kept by each group or each individual.

Some explorations may go beyond teacher-directed discussions and become small-group or individual projects. Written or oral reports should be made in these cases to share findings with the class.

Suggested discussion topics

The suggestions that follow are grouped roughly by concept. Most of these ideas can be adapted to any age level. The ideas are presented in the form of games, investigations, or questions to be posed by the classroom teacher. You and your children could spend one or more periods exploring any one of the following ideas. Some likely student responses are given, but they should be considered only as possibilities. Listen to your students' ideas and pursue them rather than your own preconceptions.

Likenesses and differences

Questions can be used to get discussions started. *Can you find two triangles that are alike in some way? Explain.* The responses could include same angles or "corners," same shapes (similar), sides the same length, two sides equal, and so on. *How many other triangles have that property? Can you draw a different triangle that has the same property?*

Can you find two triangles that are very different? How are they different? Try to find two triangles that are differ-

Table 1

Dimensions of Triangles in the Basic Set

Triangle shape #	∠A	$\overline{AB}$			$\overline{AC}$			$\overline{BC}$		
		S	M	L	S	M	L	S	M	L
1	60°	10	15	20	10	15	20	10	15	20
2	30°	10	20	25	10	20	25	6.1	10.3	12.8
3	90°	10	15	20	10	15	20	14.1	21.2	28.3
4	60°	10	20	30	5	10	15	8.7	17.3	26
5	45°	15	20	30	4.5	6	9	12.2	16.3	24.5

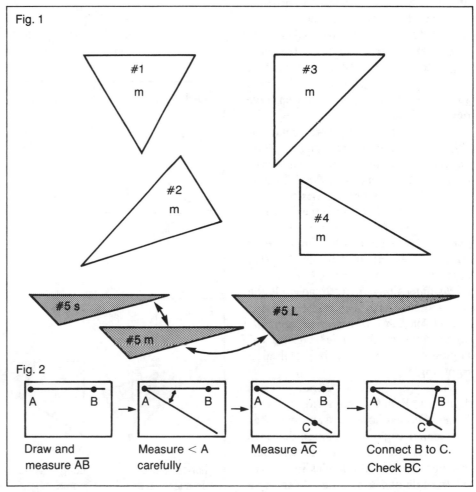

Fig. 1

Fig. 2

Draw and measure $\overline{AB}$ → Measure < A carefully → Measure $\overline{AC}$ → Connect B to C. Check $\overline{BC}$

ent in three ways. The responses could include different size, different shape, no "matching" sides.

Can you find a way to sort the triangles into two sets? Explain how you did this. Draw a new triangle that fits in each set. (Repeat for three sets.)

Games can also be used:

Game one. The first player selects a triangle. The second player selects another triangle and names some property of it that is like the first player's triangle. The next player finds a triangle with a likeness to the second player's triangle, and so on. The last player who can play is the

winner. The game can also be played by finding differences.

Game two. Each team of two players is given the same pair of triangles but of different colors. In three minutes they write down as many properties as they can find that the two triangles have in common. The longest list wins.

The triangles can also be used for hands-on measuring activities. For example, have children draw really big triangles on the floor or chalkboard that are the same shape as one of the large triangles. On the floor, masking tape can be used to extend two sides of a triangle. Have the children make each side of the large triangle a multiple of (5 times or 10 times) the sides of the original triangle. (See fig. 3)

Angles and measuring angles

If they have not already done so, have the children figure out how the #3 and #4 triangles are different from the rest of the triangles. They should discover that one corner or angle of these triangles is like the corner of a square—the triangles have a "square corner." Square-corner angles are called right angles. Have the children use these triangles to find other "square corners" in the room. We truly live in a right-angle society, a fact that many children have never noticed.

Children can investigate relationships among the different angles. *What happens when you put two square corners together on a flat surface?* By putting two right triangle shapes together, they can see that the edges form a straight line. *What happens when you put together two angles that are not right angles?* The line is "crooked." *What happens if you put four right-angle triangle shapes together?* "It goes all the way around." (Fig. 4)

Can you find a triangle with an angle bigger than a right angle? The #5 triangles have one angle larger than a right angle.

Display a #1 and a #2 triangle as in figure 5a. *Which of these two angles is larger?* Discuss how the #2 triangle has the smaller angle even though the sides of the triangle are longer. In a similar manner compare the angles in a small #2 triangle and a medium #4 triangle (fig. 5b). Try other comparisons. Children should learn that angle size is not related to the length of the sides of the angle.

What is the smallest angle in the set of triangles? The small angle in #5. Have children use it to measure the other angles in each triangle. *How many of these little angles "fit" in a larger angle?* (Fig. 6) Note that a right angle is six of these little angles. All of the angles in the set of triangles can be measured evenly with the little angle in #5. Have the children try this using the S, M, and L versions of the triangles. Use other angles as angle-measuring units, too.

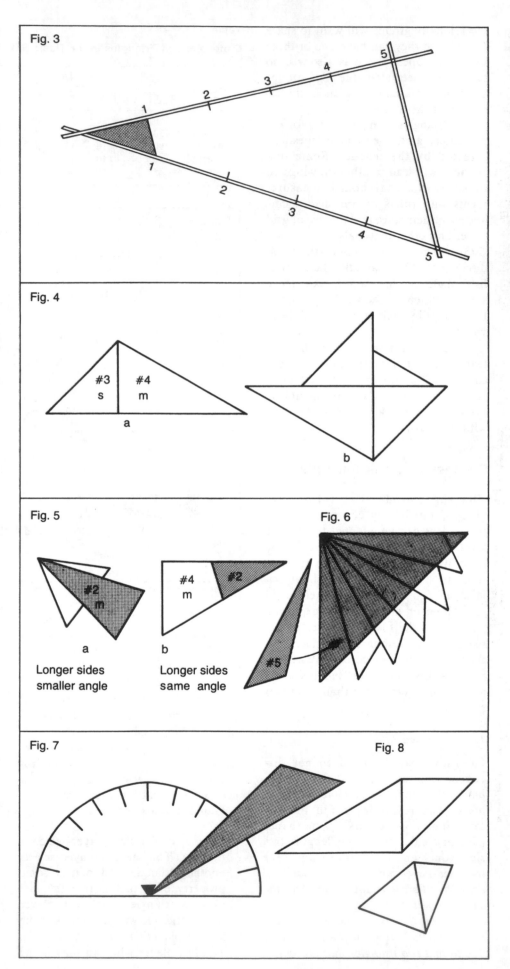

Fig. 3

Fig. 4

Fig. 5   Fig. 6

Longer sides smaller angle

Longer sides same angle

Fig. 7   Fig. 8

How many little angles are in the total of all three angles of a triangle? Twelve for every triangle. Have the children experiment with other triangles that they themselves draw. In some cases, the children may need to make estimates, using fractions for some angles.

Make a little tagboard protractor that measures in the little angles of the #5 triangle. The protractor markings can be placed on the edge of the tagboard semicircle by using the #5 triangle (fig. 7).

What is a degree? A very tiny angle. The little angle in #5 measures 15 degrees. *How many degrees in a right angle?* How many degrees would it take to go "all the way around a point?"

From here one can easily leave the triangles and go into a unit on angle measurement.

Quadrilaterals

If one triangle has an edge that is the same length as the edge of another triangle, the two can be joined to form a new figure. Most of these new figures will have four sides—a quadrilateral.

Have the children make some quadrilateral shapes from the set of triangle shapes. *How many quadrilateral shapes can you make?* Trace around some of them and then cut them out (fig. 8). There are lots of combinations. *Can the class find them all?*

Now use two complete sets of triangles. Joining two congruent ("exactly the same") triangles will form quadrilaterals with special properties. Have the class make a full set of these special quadrilaterals out of poster board. There are sixteen quadrilaterals and three new triangles that can be made for each size, S, M, and L. Figure 9 shows all of those made from #2 and #4 triangle shapes of one size.

Have the children separate all the new quadrilaterals into sets. *How many different collections of these special quadrilaterals can you make?* Possible "rules" include the following:

Same (i.e., similar) shapes made with the three sizes of the triangle shapes

Squares and nonsquares

Opposite sides go the same way (i.e.,

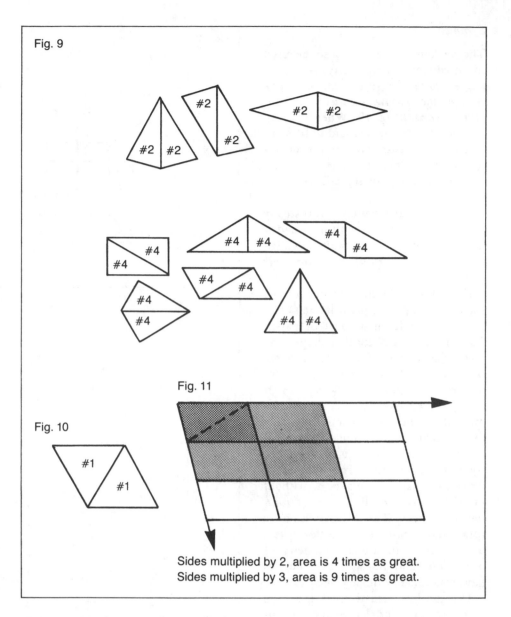

Fig. 9

Fig. 10

Fig. 11

Sides multiplied by 2, area is 4 times as great.
Sides multiplied by 3, area is 9 times as great.

the sides are parallel; the figure is a parallelogram)

All sides the same length (i.e., a rhombus, which includes squares)

All corners square (i.e., rectangles, which includes squares)

Concave and convex quadrilaterals

After all types of quadrilaterals have been named and the properties discussed, play "search and find" games. For example, *Find a rhombus that's a rectangle. Find a parallelogram that's not a rectangle.*

You can also have children select a shape and see how many names it has. For example, the shape in figure 10 is a rhombus, a quadrilateral, and a parallelogram.

Many additional investigations can be made with these special quadrilaterals.

Draw and measure the diagonals in the various figures.

What can you tell us about diagonals?

They bisect each other. They also separate the quadrilaterals into halves and fourths.

Measure perimeters.

Draw new shapes that belong to each of the different categories named earlier.

Measure the angles in a quadrilateral. What did you find? The sum of the measures of the angles of a quadrilateral is twice the sum of the measures of the angles of a triangle.

Draw larger quadrilaterals that are similar (i.e., the same shape but different in size). Make all the angles the same measure and make each side a multiple of the original side. *How much bigger is the new shape?* (Fig. 11)

Symmetry

The various shapes can also be used for discussions on symmetry. *Which triangles could be separated, with one straight line, into two identical (i.e., congruent) parts?* (Fig. 12) When this can be done, the triangle is said to be symmetrical. The two parts are mirror images of each other. The line is called a line of symmetry, or sometimes a mirror line.

You can have children investigate the lines of symmetry with the quadrilateral shapes. *How many lines of symmetry do the various quadrilaterals have?* (Fig. 13)

Have the children take one of the shapes (triangles or quadrilaterals) and trace around it on a sheet of paper. Call this the "box" for that shape. *Can your shape be placed in its box in more than one way?* Of the triangles, only #1 will fit more than one way without being turned over. A shape that fits more than one way without turning it over is rotated to get from one fitting to the next. Such a shape is said to have rotational symmetry. The number of ways the shape fits its box (without flipping it over) is the *order of rotational symmetry*. The #1 triangles have rotational symmetry of order three. Squares have rotational symmetry of order four and nonsquare rectangles have rotational symmetry of order two.

Use the box idea to investigate the symmetry of other shapes that the students draw. The children should discover that a shape must have either line or rotational symmetry if it is to fit in its box in more than one way.

Solids

Solid figures can be generated with the various shapes. Have the children hold a #3 triangle so that one of its shorter sides is vertical. *What would you see if the triangle started spinning around this edge?* A cone. *What would you see if the triangle started spinning around the longest side?* (See fig. 14)

Find a triangle with a line of symmetry. *What would you get if you spin the triangle on its line of symmetry?*

Try spinning other shapes, especially the quadrilaterals. *What solids can you "make"?*

Select a solid shape. *Can you think of*

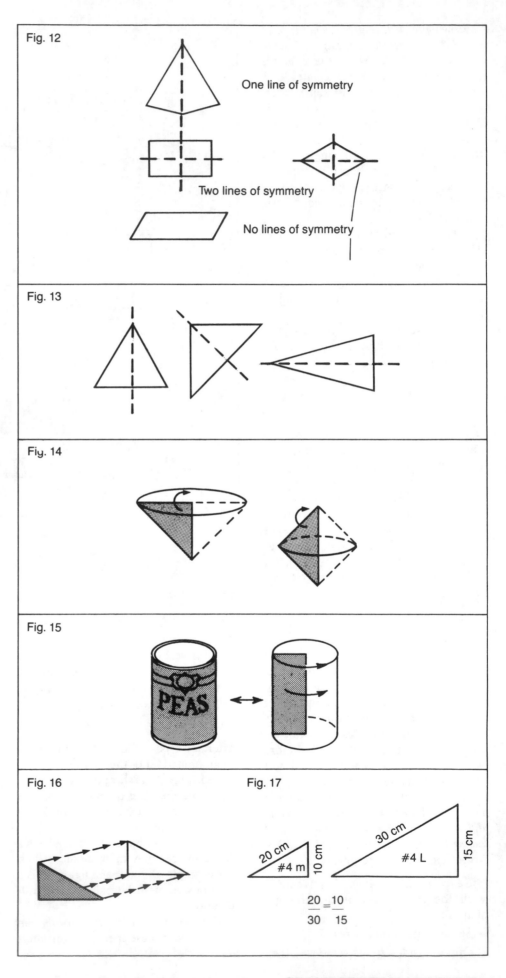

Fig. 12

One line of symmetry

Two lines of symmetry

No lines of symmetry

Fig. 13

Fig. 14

Fig. 15

PEAS

Fig. 16   Fig. 17

20 cm   10 cm   30 cm   #4 L   15 cm   #4 m

$$\frac{20}{30} = \frac{10}{15}$$

a shape that could be used to "spin it"? (Fig. 15)

Instead of spinning a shape to make a solid, have the children try "sliding" a shape to make a solid. (Fig. 16) Again, the triangles are only the beginning. Spin or slide various other shapes—some odd, some regular. Note that a circle will "spin a sphere" and "slide a cylinder." A rectangle "spins a cylinder" and "slides a box."

Ratios and similarity

Triangles #1 and #4 have been designed so that at least two edges are a whole number length in centimeters. Students studying metric measurement and ratio can compare the lengths of corresponding sides or similar triangles. The ratios of these sides will, of course, be the same. You can extend this project to very large shapes drawn on the floor (see fig. 3) or even drawn on the playground (measured in meters).

In figure 17, the corresponding sides of the two similar triangles are in a ratio of 20 to 30 or 2 to 3. The ratios of sides within a triangle can also be compared. In figure 17, for example, the ratio of the shorter to the longer side is 10 to 20 (for the #4M triangle) and 15 to 30 (for the #4L triangle), each of which is equivalent to 1 to 2.

The ratio of areas of two similar triangles is also interesting. To observe this, it is best to select a triangle and construct a larger, similar triangle by doubling or tripling each side. If this is done by tracing around the original triangle, the comparisons of areas are immediately obvious, as in figure 18. Some students will be able to discover a rule: If the sides of a figures are multiplied by n, then the area will be multiplied by n^2. This works with all plane figures.

Tesselations

Have the children select one of the small or medium-sized triangles and make forty to fifty copies of it from two colors of construction paper. These can then be used to put together a tiling pattern or tesselation. Shape #1 is the easiest for young children to use. Shapes #3 and #4 are also easy. All triangles, however, will tesselate or fill up a surface with no gaps (fig. 19).

Tesselation art opens up a whole new world of geometry beyond the scope of this article. However, since every triangle and every quadrilateral can be used to make tiling patterns, the fifteen triangles and the various quadrilaterals made from them are a great place to start. (See also "Let's Do It" in the November 1980 issue for more ideas about tesselations.)

Fig. 18

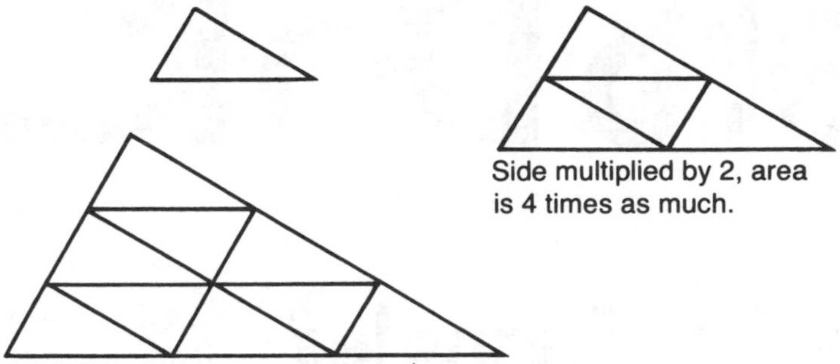

Side multiplied by 2, area is 4 times as much.

Side multiplied by 3, area is 9 times as much.

Fig. 19

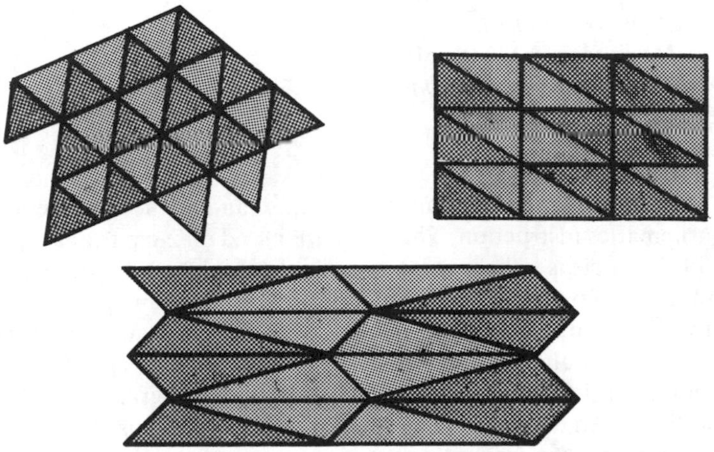

Notes on the Drawings

1. All triangles in all of these drawings are a scale version of one of the five basic triangles described in the article. The triangles are described by their angles here. The actual side lengths are in table 1.

   #1: 60°-60°-60°   #4: 30°-60°-90°
   #2: 30°-75°-75°   #5: 15°-45°-120°
   #3: 45°-45°-90°

2. Whenever two triangles of the same shape, even though different sizes, are shown in one figure, the color of these triangles should be the same. Color is not required, however.

Conclusion

We have found that when children work with these triangle shapes the types and levels of real investigations that grow out of the activities are wide and varied. The only real rule for working with these materials is to listen to the children and help them learn to explore by selective questioning. You can (and will) learn as they do. ◉

Let's Do It

Let's Try Triangles

By **Marcia E. Dana**
Orchard Ridge Elementary School
Madison, Wisconsin
and **Mary Montgomery Lindquist**
National College of Education
Evanston, Illinois

Skill in problem solving is a primary goal of mathematics instruction. The purpose of this article is to help you build a collection of problems that can be posed to children of varying ages or abilities. These particular problems require searching for different possibilities that fit a given condition. The number of solutions that are found and how they are found will vary with the groups of children involved. In fact, for some children you may want to use these searches as class activities rather than as projects for individual students.

One outcome of searches such as the ones described here is that children may become more systematic in their approaches to problem solving. However, before they arrive at the point of systematic searching, children need to understand the problem. And understanding a problem often necessitates some "playing around" with the conditions. The "playing around" in itself can produce unexpected outcomes as well as a better understanding of the shapes used in these searches. Remember that the process of analyzing a problem in an orderly manner is not easy—do not expect major leaps by all children.

The activities that follow involve the use of triangle paper. Equilateral triangles, like squares, can be arranged to cover a sheet of paper and such "graph" paper has a great variety of uses. (Triangles that measure 2 cm on each side are a good size for triangle

paper, and all the activities described are based on 2-cm triangle paper.) Triangle graph paper is quite different from square graph paper, and many of the things that can be done with it are intriguing because it is less familiar than the usual graph paper. You and the children can use triangle paper with a variety of activities, only some of which are described here.

While children are doing these activities, they are learning about, or developing readiness for, the following mathematical processes and concepts:

1. problem solving
2. collecting and organizing data
3. looking for and recognizing patterns
4. making and testing hypotheses (predicting what will happen)
5. symmetry
6. triangles (what a triangle is and is not), line segments, and angles
7. perimeter and area
8. congruency

Children are also enjoying the challenge of puzzles or problems and the satisfaction of solving one or the other through their own efforts.

Putting Triangle Puzzles Together

Young children can gain experience with the properties of triangles by putting puzzle pieces together to make a triangle. You can make these triangle puzzles as easy or as hard as you wish.

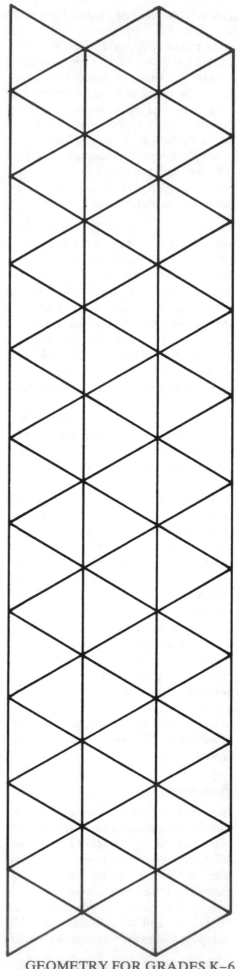

Preparing the puzzles

1. Trace around a large triangle on a sheet of triangle paper. A good size is 10 small triangles (20 cm) on each side. Cut out the large triangle, then use it as a pattern to make your triangle puzzles. You will need two large triangles for each puzzle.

2. For each puzzle, use a standard size sheet of construction paper, a different color for each puzzle. Trace the triangle pattern twice on each sheet of construction paper, then cut the two triangles out. Cut *one* of the two triangles into pieces (fig. 1). Leave the other triangle whole so the children can put the pieces of the cut-up triangle on it as they try to put the cut-up triangle together.

3. Make as many triangle puzzles as you wish. You could store each pair of triangles in a large envelope for easy use.

Using the puzzles

The following are the properties the children will be investigating as they do the puzzles:

A triangle has three straight sides.
A triangle has three corners (angles).
A triangle is not curved.

Once the children have put together several of the puzzles, you could ask them to try putting the puzzles together without using the second triangle of each pair as a help. Another possibility would be to give each child an uncut triangle and let them cut their triangles into from two to six pieces. Children could then trade puzzles with friends and try to put together the friend's puzzle. (You might want to try them, too.)

Searching for Shapes Made with Triangles

A systematic search for figures made with certain numbers of triangles can be conducted; the results of the search can be recorded on triangle paper. This type of search has several benefits. One important benefit is that children can learn that when they are stuck on a problem or puzzle they can try a different technique or pattern. Another benefit is that children will see that some kind of system will usually get them

farther than trial and error will. One more idea that is involved here is that of using what you already know (in this case, shapes made with fewer triangles) to help solve your problem (in this case, finding shapes made with more triangles).

Most children will probably need to use triangles that they can pick up and rearrange easily. To make such triangles, trace around four of the tri-

Fig. 1

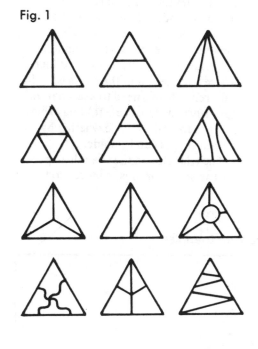

Fig. 2

(4 cm on each side)

Fig. 3

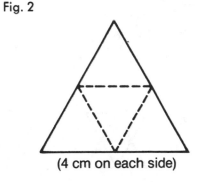

not this   or this

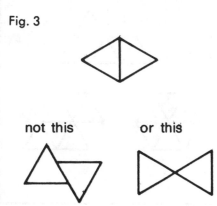

angles on the triangle paper (fig. 2), but do not draw the inner lines. Make tagboard triangles the size of the triangle you have traced. Triangles of this size will be large enough for most children to handle. Each child who is trying this activity will need six large triangles. If you do the activity with only a few children at a time, you will not need as many triangles.

Before beginning the search with the children you will need to introduce the rule they are to follow when putting the triangles together: *The triangles must always be put together so that their sides coincide* (fig. 3).

Begin with shapes made by two triangles. Every child will easily find the only shape that can be made with two triangles. Each child can record the two-triangle shape on a sheet of triangle paper by shading two small triangles next to each other. Next they try shapes made with three triangles—again there is only one—and record the result. Be sure the children are convinced that there are no other different shapes that can be made with three triangles when they follow the rule. You will also need to tell them that one shape is not different from another shape if you can move it and match it with the other—in other words, if the shapes are congruent. Children can check to see if shapes are the same (i.e., congruent) by cutting out a shape they have shaded on their triangle paper and trying to match it with another on the paper. This same test can be used with any of the larger shapes, too.

Next have the children try to make shapes from four triangles. Do not tell them, at first, how many four-triangle shapes there are; just let them try to make shapes with four triangles for a while. Not every child, however, will be able to find all three four-triangle shapes by himself or herself. You may begin helping children by guiding them to systematize their search. Suggest that they make the three-triangle shape and then put one more triangle on it, in first one place and then in another. This same technique can be used to find five-triangle shapes or even six-triangle shapes (fig. 4). As the number of triangles in the shapes increases, more and more testing will probably be needed, and more children will be unable to find all the possible shapes. And it may not be obvious to some children that every five-triangle shape will be the same as one of the four five-triangle shapes in figure 4.

In finding the different possible shapes for five and then six triangles, another technique that helps is counting how many triangles are in a line in a shape (fig. 5).

Searching for shapes made with triangles is a project that will probably take more than one day, but the sense of accomplishment the children will feel will be worth the time expended. You might want to post on the bulletin board all the shapes found by the class. Or you might make and duplicate a sheet showing and telling about the shapes. Copies of the sheets could be handed out to other classes or taken home to show parents. As a real problem for some or all of your class, you can challenge them to find as many seven-triangle shapes as they can.

Folding-Triangle Shapes

Strips of equilateral triangles can be folded along the lines to make a variety of shapes, including many of those in figure 4. The children will be used to folding rectangular figures in which the two folded parts match when folded. It will be a good experience for them to see that folding does not necessarily result in matching overlapping parts. Folding that involves 60-degree angles encourages the children to learn through hands-on experience and to predict and experiment by asking themselves questions: How can I get two triangles over here? If I fold there, where will the rest of the strips go? The children will also need to use what they have learned from these folding experiences to logically decide whether or not certain shapes can be made.

The triangles on the triangle paper are too small for most children to fold.

Larger triangles, the size of four of the small triangles, will be best. Using the triangle paper as a pattern, trace six of the large triangles end to end (fig. 6). Make copies of the six-triangle strip. (The triangle strips will work better if they are duplicated on tagboard or construction paper.) Give each child a copy of a triangle strip. Have them cut out their strips and carefully crease the strips along the five inner lines.

To familiarize children with the way these triangle strips fold, ask the children to fold one triangle behind so they can see only five triangles in a row. Then have them fold the strip so that only four triangles in a row are showing. Repeat this so that three, then two, and finally only one triangle is showing.

Next draw the shapes in figure 7 on a chalkboard or trace them on triangle paper (the triangles need not be the same size as the ones the children are

Fig. 4   Shapes made with triangles

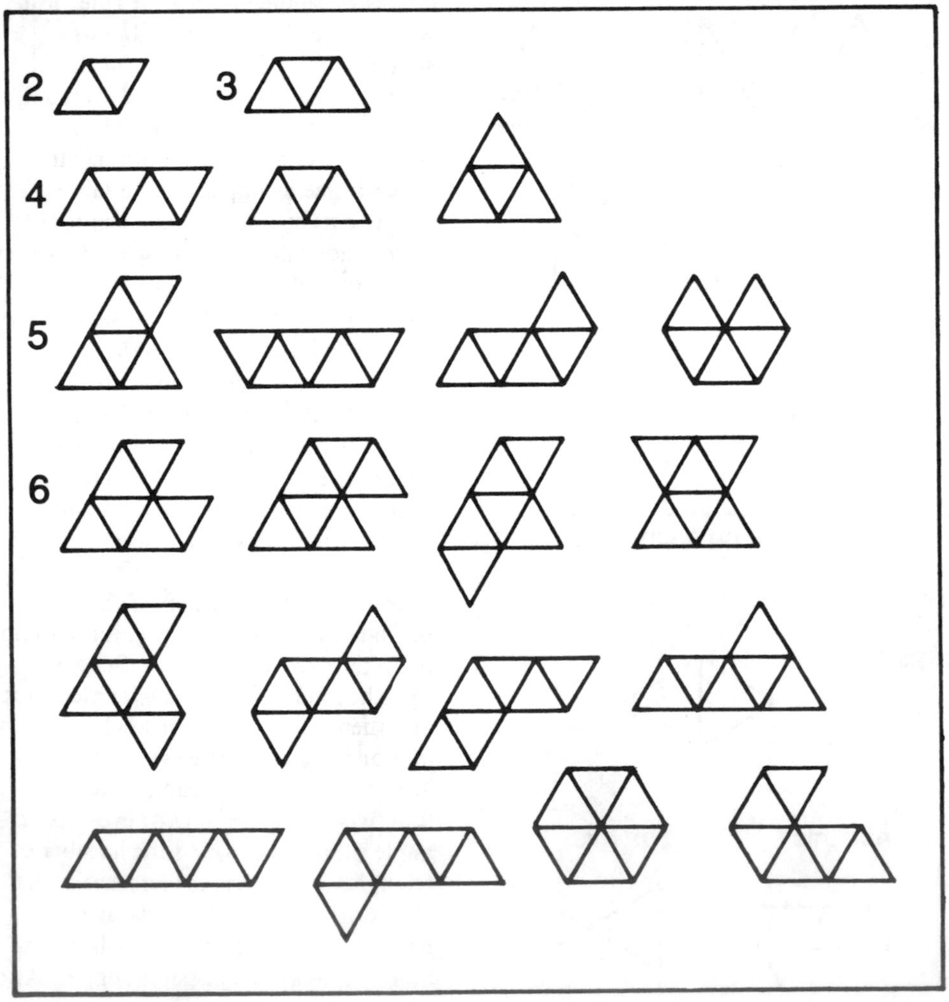

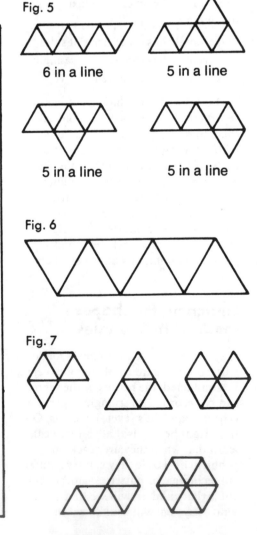

Fig. 5

6 in a line   5 in a line

5 in a line   5 in a line

Fig. 6

Fig. 7

GEOMETRY FOR GRADES K–6

using). Ask the children to fold their triangle strips to try to make each of the shapes. (The second and fifth shapes cannot be made from a six-triangle strip.)

Once the children have had experience with a six-triangle strip, they can try folding strips of seven or eight triangles—you can make them in the same way you made the six-triangle strip. Make copies of the shapes in figure 8, then have the children try to fold the seven-triangle, and later the eight-triangle, strips to make the shapes. Some of the shapes in figure 8 cannot be made with either strip, which makes the activity more challenging. You may want to do this particular activity as a class project and give the children a week in which to find all the shapes that can be made with strips of seven and eight triangles.

If the children really enjoy folding triangle strips to make shapes, they may want to try folding other shapes made of triangles. The shapes in figure 9 are possibilities. After the shapes in figure 9 have been cut out, have the children cut along the red lines. Let the children experiment with these shapes and find out for themselves what shapes can be made. The shapes in figure 9 can be used to make some of the shapes in figure 4 that could not be made by folding triangle strips.

Fig. 8

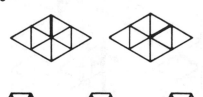

Fig. 9

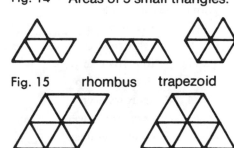

Minisearches

The searches suggested here are built around one of two ideas: How many of one shape fit inside a given shape? or How many different shapes with a given property can be found within the given shape? Some of the searches described here can be done by a more or less trial-and-error procedure. A complete solution, however, is often reached more quickly by a systematic approach.

First minisearch

How many four-triangle parallelograms are in a nine-triangle triangle (fig. 10)? Have children shade in the possibilities on copies of the nine-triangle triangle. It may help some children to cut out a copy of the parallelogram and move it around on the triangle. The solution is shown in figure 11.

Second minisearch

How many two-triangle rhombuses are in a nine-triangle triangle (fig. 12)? How many three-triangle trapezoids? How many five-triangle trapezoids? Respectively, there are nine two-triangle rhombuses, twelve three-triangle trapezoids, and three five-triangle trapezoids in the nine-triangle triangle.

Third minisearch

How many shapes with a perimeter of 6 units are in the nine-triangle triangle? Use the length of a side of a small triangle as a unit. The solution is shown in figure 13.

Fourth minisearch

How many shapes with an area of five small triangles are in the nine-triangle triangle? The solution is shown in figure 14.

You can vary the searches by using another shape in place of the nine-triangle triangle. A rhombus or a trapezoid made from eight small triangles (fig. 15) are reasonable alternatives. Comparing the results found in using these two shapes could be worthwhile.

Triangle Paper Searches

In these searches the children examine a whole page of triangle paper to see how many different sizes of a certain shape they can find. For example, they might search to see how many different sizes of regular hexagons there

Fig. 10

How many
are in

Fig. 11

Fig. 12

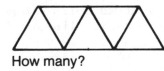

How many?   How many?

How many?

Fig. 13   Perimeters of 6 units.

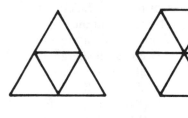

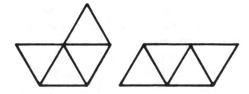

Fig. 14   Areas of 5 small triangles.

Fig. 15   rhombus   trapezoid

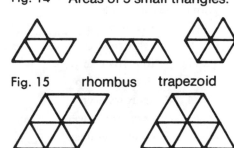

are on a page of triangle paper. (There are five.) This type of search should encourage a child to see a need for being organized. Some children will naturally approach a question like this systematically, but many will use a random, trial-and-error approach. Let children try their own ways, then discuss their approaches. If no one suggests an organized approach, you might suggest that they look for a hexagon with sides of one unit (one side of a triangle), with sides of two units, of three units, of four units, and so on.

First triangle paper search

How many different sizes of equilateral triangles are on a sheet of triangle paper? How many different sizes of rhombuses? (There are eleven different equilateral triangles and eight different rhombuses.)

Second triangle paper search

How many different sizes of parallelograms are on a sheet of triangle paper?

In the second triangle search, some children will profit from cutting out parallelograms and sorting them. Once they begin sorting, they may come up with an organized approach. One systematic approach would be to see how many parallelograms are "one row deep" (fig. 16). In "one row deep," the parallelograms are made from an even number of triangles, from two to twenty-four, or twelve parallelograms all together. Next look at the two-row deep parallelograms. Be careful of the first two-row deep parallelogram in figure 16. This parallelogram is the same size as the second one in the one-row deep parallelograms. Next look for parallelograms that are three rows deep, four rows deep, five rows deep, and so on. But also look to see that no parallelogram is the same size as another parallelogram (that is, that no two parallelograms are congruent). In this manner you should find, in addition to the twelve one-row-deep parallelograms, the following:

2	two rows deep
9	three rows deep
7	four rows deep
6	five rows deep
4	six rows deep
3	seven rows deep
1	eight rows deep

Another systematic approach to the question of how many parallelograms would be to classify parallelograms by area—two triangles, four triangles, six triangles, and so on.

Third triangle paper search

How many different trapezoids are on a sheet of triangle paper? (There are sixty-six.)

Symmetry Searches

In these searches children create designs as they explore how changing the beginning of the design will or will not change the entire design. The triangle paper provides not only a way to examine figures with one or more than one line of symmetry, but also experiences with lines of symmetry that are diagonal lines. Using a diagonal line as a line of symmetry is more difficult than using a horizontal or vertical line so symmetry searches on triangle paper are best done with older children. It might also be beneficial for your children to try making some designs on triangle

paper before they do the searches. For example, they might do the following:

1. *Mark a diagonal line as the line of symmetry and color in five triangles on one side of the line of symmetry. Then color the images of the triangles on the other side of the line of symmetry so that the design is symmetric (fig. 17).*

Some children will find it easier to turn the paper so that the diagonal line is vertical for them. Others will need to fold the paper on the diagonal line and punch through the colored triangle to the one it matches. Opening the paper, they then can color in the corresponding triangle.

2. *Mark as lines of symmetry a diagonal line and a vertical line that crosses it. Color in one triangle. Complete the design so that it will be symmetric with respect to both lines of symmetry (fig. 18).*

This procedure, because of the structure of the paper, will automatically produce a third line of symmetry. Let the children discover this. Again, some

Fig. 16

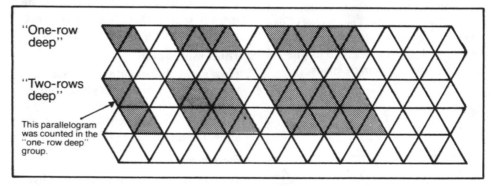

Fig. 17

Fig. 18

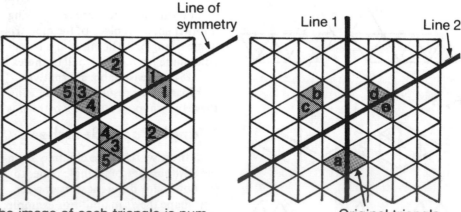

The image of each triangle is numbered the same as the triangle.

Fig. 19

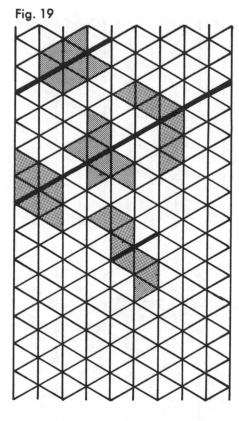

Fig. 20

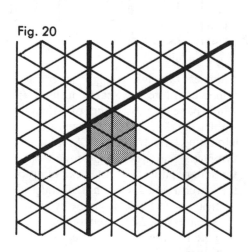

Fig. 21

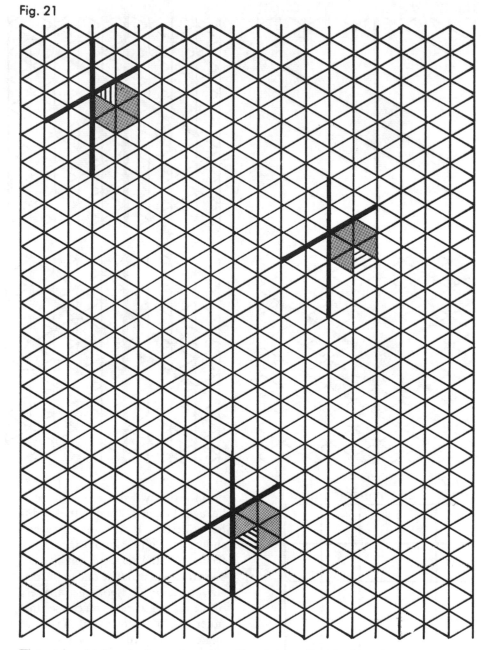

The striped triangle in each case will produce the design shown.

children will need to fold on each line of symmetry and punch holes. Make certain that the children see that each time they color in a triangle they need to consider the image of this new triangle with respect to each line of symmetry. For example, in figure 18, triangle *a* is the image of the original triangle with respect to line 1, triangle *b* is the image of the original triangle with respect to line 2, triangle *c* is the image of triangle *a* with respect to line 2, triangle *d* is the image of triangle *b* with respect to line 1, and triangle *e* is the image of *c* with respect to line 1.

After the children have become fa-miliar with the triangle paper and sym-metry, let them try one of the following searches. Also, encourage children to invent their own searches.

First symmetry search

Mark a diagonal line as a line of sym-metry. Color four triangles so that (1) one side of the triangle touches the line of symmetry and (2) each triangle touches at least one side of one of the other three triangles. Complete the de-sign so that it is symmetric. How many different designs can you make? (There are five, as shown in figure 19.)

Second symmetry search

Mark a diagonal line and a vertical line that crosses the diagonal line as lines of symmetry. Find the regular hexagon made from six triangles and bounded by these two lines (fig. 20). Color one triangle in the hexagon red. Now com-plete the design with another color so that the design is symmetric with re-spect to both lines. How many different designs can you make? (There are three designs, as shown in figure 21.)

Third symmetry search

Repeat the second search, but this time

Fig. 22

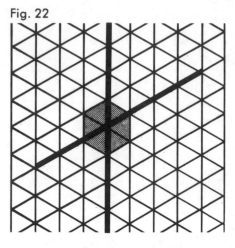
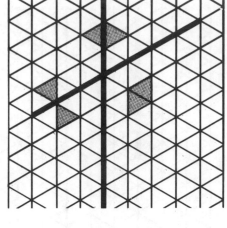
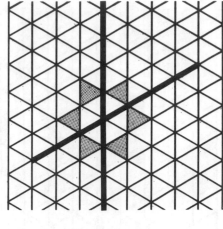
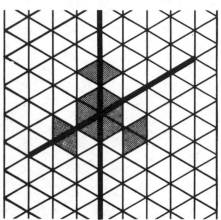
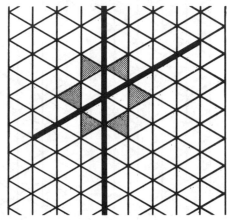
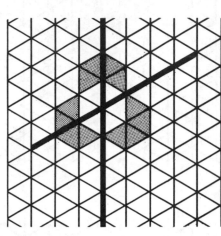

Fig. 23

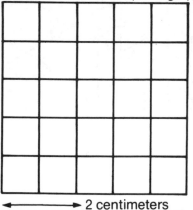

1 centimeter (or larger)

2 centimeters

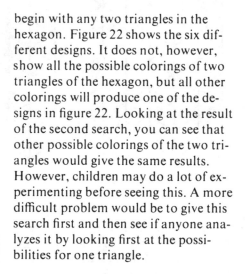

begin with any two triangles in the hexagon. Figure 22 shows the six different designs. It does not, however, show all the possible colorings of two triangles of the hexagon, but all other colorings will produce one of the designs in figure 22. Looking at the result of the second search, you can see that other possible colorings of the two triangles would give the same results. However, children may do a lot of experimenting before seeing this. A more difficult problem would be to give this search first and then see if anyone analyzes it by looking first at the possibilities for one triangle.

Summary

In all these searches, triangle paper has been used. You may vary the searches by using a different type of paper such as square paper or right triangle paper (fig. 23). In general, similar searches are easier with the square paper and more difficult with the right triangle paper. The minisearches could be done with a square made from 9 small squares or a right triangle made from 8 small right triangles. The right triangle paper also makes an interesting search of the whole page for the number of different right triangles, rhombuses, parallelograms, and trapezoids. Similar symmetry searches may be done with the square paper. For ones like symmetry search 1 and 2, use a horizontal and vertical line as lines of symmetry and the squares in a 2-by-3 rectangle bounded by these lines. A diagonal line and a vertical line as lines of symmetry on the right triangle paper will produce designs with four lines of symmetry. □

GEOMETRY FOR GRADES K–6

Blocks

2

Every child should have opportunities to play with blocks. A set of blocks can consist of matched pieces or a miscellaneous collection of solids. A teacher can even assemble a set of "blocks" from empty shoe boxes or cereal boxes.

Blocks can be an avenue of mathematical experiences for very young children. In their play with blocks, children intuitively learn to recognize square corners (right angles) and to match shapes. Blocks also introduce children to the notion of three-dimensional shapes.

In "Experiences with Blocks in Kindergarten," Liedtke describes numerous ways of using blocks with young children. He also relates individual activities to specific mathematical concepts.

Bruni and Silverman ("Using Cubes") use a matched set of cubes to develop mathematical experiences, including counting and applying the notions of addition and subtraction. The same authors ("From Blocks and Model Making to Ratio and Proportion") relate the experiences of young children with blocks to the subsequent teaching of specific mathematical topics.

These three articles show that blocks are not only good playthings but also worthwhile instructional aids.

Experiences with blocks in kindergarten

W. LIEDTKE

An associate professor at the University of Victoria in British Columbia, Werner Liedtke teaches mathematics education courses to undergraduate students who want to teach in grades K–3. He also conducts seminars on teaching and supervises student teaching.

The availability of a great variety of manipulative materials in kindergartens today makes it relatively easy to capitalize on a young child's eagerness to explore and interpret the world around him. A set of blocks is just one such manipulative aid which is well suited for the provision of preschool experiences in mathematics. A variety of experiences based on the use of blocks are explored in the following paragraphs by presenting some sample problems. These problems could be posed to individuals or to small groups of children as they are engaged in free play. Some of the problems may also be suitable for large group presentations and discussions.

Learning from their own play and from the imitation of other children and adults, it is not long before young children will recognize similarities, differences, detailed characteristics, and properties of the blocks. They will experience new ways of looking at objects in their environment, gain confidence in solving problems, and recognize or even label many common geometrical shapes. Many of the contacts with the blocks should be informal and the structured activities should probably be introduced as casually as possible. Observations of the way young children interact with the blocks in their free play is essential if the activities outlined here are to be introduced effectively.

A good way to begin is to provide each child with a number of blocks and let him do whatever he wishes. As the child is playing or building, he should be encouraged to show and talk about what he has made. Objects from his immediate environment will likely be presented by various ingenious constructions. To relate the objects to the child's experience, various questions could be posed to individuals or groups of children: "Pick up any block. Tell us what it reminds you of. Why?" "What would you call this block? Do you know another name for it?" Or more specifically, "Pick up a block that looks like something you could find in your kitchen (garage, basement) or at the playground or store, and we'll try to guess what you are thinking of." If the members of the group cannot guess what the child has in mind, he could be asked to give some more hints. Giving such hints without supplying the answers is, for the first few times at least, a rather difficult (and sometimes amusing) task for young children.

An activity that will result in quick actions and responses consists of simply naming an object (TV, ice cream, tool, house, tent, grocery store) or an action (playing, sleeping, skating, surfing) and having a group of children select from a pile of blocks the one that they think can be associated with that object or action. Children could then be asked to tell the group what they were thinking of as they made their selection. Following this, one of the children in the group could be asked to look at a pile of blocks, name an object or action, and then have the other members of the group try to pick out the one specific block he was thinking of. The first child to identify the correct block could then be "it" and would get a turn to name an object or action.

The problem of finding a block that is "just like" one that has been selected for them gets the children to think of shapes that are the same. To have children consider similar shapes, a block can be presented to them and they can be asked to find, from a small pile of blocks that does not include the one shown to them, a block most like it.

Children's responses to the problem of being presented a handful of blocks and then being asked to select all the blocks that are in some way the same will give an indication of how aware the children are of similarities and differences in considering some of the common characteristics displayed by the blocks. Two very simple activities or games can be used to help children become more aware of the parts of the blocks. In the first, hidden from the child's view, a block is placed in a bag. The child is then given the bag and asked to find a block that is either "just like it" or "most like the one in the bag." In the second, as children sit in a circle, facing the blocks, they are asked to hold one hand behind their backs. One child then places a block in the hand behind the back. Without looking at the block they have been given, the children are then to solve the same problem(s) as for the bag activity.

There is another way of helping children discover the similarities and differences between blocks. A block is chosen—

a 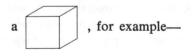 , for example—

and the children are asked to name it or to give it a name. If, for example, the children agree to call it a cube (any other name would do, as long as everyone agreed), then various questions or problems could be presented: "If this is a cube, what name should or could we give to this part?"

"Can you find other blocks that have just as many of 'these'?" "Can you find some with more?" "Can you find some with less?" "Which block in this pile has the most or the fewest?" "Can you show me with your fingers (preferably without referring to number) how many of 'these' the cube has?" A similar procedure can then be followed for discussing the other parts of the cube—

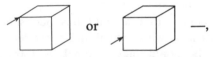

and in this way new classification schemes can be introduced. *More, fewer,* or *same number of* faces; *more, fewer,* (one? or none?) or *same number of* edges; and *more, fewer,* (one? or none?), or *same number of* corners can be used when sorting a given number of blocks.

To get the children to consider some of the characteristics of the blocks that they have discovered, a simple game can be devised. A few children are seated around a table and each child is given the same number of blocks. One person begins by placing one block in the middle of the table. Then each child in turn takes from his own pile of blocks a block that must be different in *one and only one* way and places it next to the one in the middle. Should rules be needed for such things as taking or missing a turn, they could be made up with the children. Perhaps more important than the rules is the opportunity given to the children to talk about their choices and then to find out why they thought their choices were correct. For another game, the "one difference" rule could simply be changed to "two differences" or "three differences."

The children can also be led to discover certain characteristics of the blocks by giving them a number of blocks that are alike or similar. The children are then asked to use these blocks to construct a building that is tall (or strong, or wide, or fancy). The same instructions are repeated for other piles of blocks that are similar to one another but different from the ones used for the previous task. The children's responses to such questions as, "Which of the blocks are better for making a tall building and why?" or "Why is one pile of blocks better for certain buildings than another?" will give an indication as to how aware the children are of the various characteristics displayed by the different piles of blocks.

Assuming that a child has had some experiences with pencil-and-paper exercises, he could take a block and trace around it. Then he could be asked to pick out some blocks that he thinks would give him a drawing that is either the same or similar to the first drawing. A discussion about a number of such drawings—

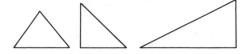

on how these are the same and how these are different, could lead to some interesting generalizations about triangles. The children could then be asked to draw a picture of something like a house, a man, or an animal using just triangles. Their pictures could be displayed and then the children could be asked to draw another picture using just rectangular shapes. An evaluation or a comparison of the pictures could be made by having the artists respond to questions: "Which picture do you like best? Why?" "Which shape do you like

to draw with? Why?" "Now, make a picture using both shapes at the same time?" For the last task, the children could be asked to provide reasons for using the shapes where they did.

The problems or tasks that have been described suggest one possible sequence for the exploration of the blocks and how they could be used to have children discover where two-dimensional shapes or special closed curves come from. The blocks as such are suitable also for various other problems and most teacher guides contain numerous suggestions. Activities that deal with prenumber ideas and measurement can be made challenging and interesting. At the same time the children's responses to these tasks will give an indication as to how ready the children are to deal with these topics.

A simple pattern can be constructed

and then the child is asked to guess which block should come next in the sequence. If the child answers correctly, the question, "How did you know?" should be posed and then he could be asked to add more blocks to the sequence in either direction. If the child's guess is incorrect, he could be challenged to try and find the correct block by trial and error. Variations for this type of activity or problem are numerous indeed. (See fig. 1.)

The questions that could be asked for each of the examples in figure 1 are "What do you think the next block (or building) looks like?" "How do you know?" "Try to put down a few more to make the row (pattern or sequence) longer."

The children should have a chance to construct patterns of their own and perhaps have a friend try to guess the secret for the sequence that they make. An interesting riddle can be created by taking a sequence and hiding one or more members of the sequence from the child's view with a piece of paper or cardboard.

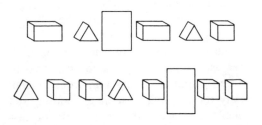

Then the following questions could be asked: "Show me with your fingers, how many blocks you think are hidden?" "Can you find a block that looks like the one-(ones) that is(are) hidden?" "How did you know?"

"Buildings," like those shown in figure 1, lend themselves to some interesting classification tasks. A group of houses is constructed and a child is asked to look at them and try to find some that are in some way alike or the same.

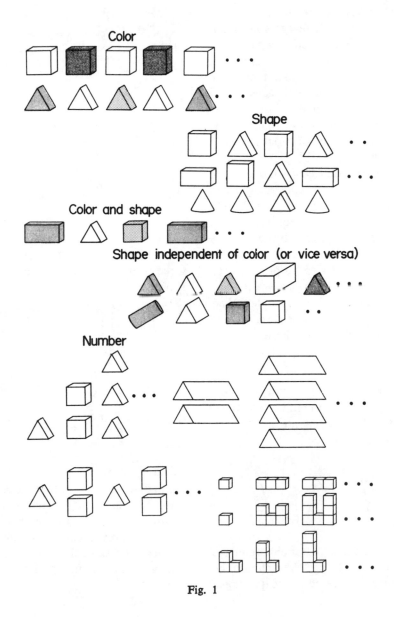

Fig. 1

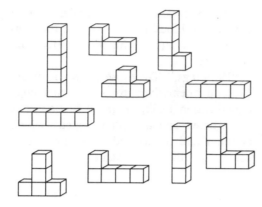

Various groupings are possible. Most children will consider such features as height or shape. A few may even recognize that various different-looking buildings have the same number of "rooms." No matter which characteristic is considered in solving the problem, numerous variations can be created by simply changing the shape of the building or the number of blocks used to construct them.

The simple experience of sharing a pile of blocks with a friend, and being asked to do this without counting, will help children realize that two equivalent sets can be established simply by using the process of one-to-one correspondence. Some activities could also be presented to give the children a chance to use one-to-many correspondence. For example, "Find enough little blocks such that there are two (or three) for each one of these"

Instead of saying "two" or "three," perhaps the desired number could be indicated to the children by showing fingers, "Try to find this many,

for each of the blocks." After children have completed the matching, they should be given a chance to respond to the questions,

"Are there more little blocks, more big blocks, or are there the same number of each? How do you know?"

The opportunities for having the children use the process of matching or one-to-one correspondence are numerous indeed. Many daily activities such as handing out pencils or paper, setting a table or distributing toys are well suited for such an activity. One simple variation of the process could consist of having children match some unusual arrangements.

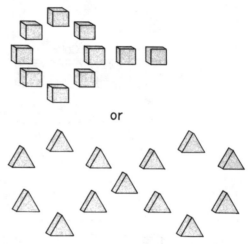

In another variation, children could be given a chance to consider or compare three or more sets of objects. They can be shown a set of blocks

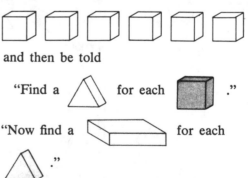

and then be told

"Find a △ for each ■."

"Now find a ▱ for each △."

No matter what the activity, the questions, "Are there more of one, more of the other, or are there the same number of each?" and "How do you know?" should be asked. In the last example, children could be asked to make another or third comparison.

Since the blocks differ in shape, size, and sometimes in color, many of the children may become aware of the fact that number is independent of these characteristics. However, most children of this age don't realize that number is independent of arrangement. They will not be able consistently to recognize equivalence after the physical arrangement of the elements of one set is altered. These children may perhaps profit from the following activities. A die, or a block with the markings of a die, is rolled and the children are asked to pick up as many blocks as the die shows. Some of the "homemade" dice could perhaps have markings different from those on a regular die.

The children are then asked to pick out several groups of blocks, with each group having as many blocks as the die shows. The children could also be challenged to make each group look different without adding or taking away any of the blocks.

Some children may be ready to choose from a series of numerals printed on a piece of cardboard the name given to the groups. They could be shown a set of numeral cards

and asked, "Which of these would you choose as a name for your groups?"

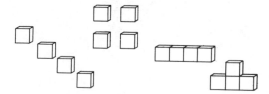

If they are not ready for this task, then the activity with the die is perhaps a good way of introducing them to the number names (four) and numerals (4).

Most sets of blocks are also suited for various ordering activities. Children can be given problems such as, "Choose a block from the box (or from in front of you) that should come next."

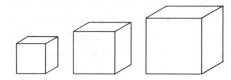

Then they should be asked, "How do you know it is the correct one?" A question like, "Where would you put this block?"

will give the children a chance to discover relationships by adding to either end of an ordered sequence or by inserting into the sequence. Children also could be asked to construct similar ordered sequences with different blocks.

Problems could also be presented that require matching and ordering. "Look at these blocks."

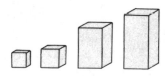

"Which of the triangles

would you match with which of these blocks?" "Why did you put 'this' one 'here'?" Children could also be challenged to find a common characteristic and order for a group of dissimilar blocks

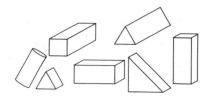

and order these according to the magnitude of that characteristic.

In another activity, a tower could be built for the children and they could be asked to build one, with different blocks and a short distance away, that is just as high. The children's responses to the questions "How do you know yours is just as high?" and "Show me in some way that yours is just as high as mine" will give some insight or indication as to how ready they are to deal with measurement. Most of the children will rely heavily on perception—"I can see"—and will see no need for the use of a unit of measure. Similarly, some may move the towers together and compare the top parts or end points. A few may actually use parts of their body to compare the heights of the towers, thus indicating that they see the need for using a unit. Having children attempt to duplicate, on a table, the height of a tower that has been built for them on a box or on the floor, will give an indication of whether or not they are able to consider both the point of departure and the point of arrival for the tower they are building.

A few suggestions for activities have been made, but the children themselves are often an excellent source for more ideas. They will learn and develop through both imitation and free play. By trying to solve the problems suggested here, and imitating an adult or another child, youngsters may get ideas they never had before. They will discover new ways of looking at things and objects around them and thus expand on the ideas and capabilities they already possess.

By **James V. Bruni** and **Helene Silverman,** Herbert H.
Lehman College, City University of New York.

Using cubes

Those wooden or plastic cubes often found in primary-grade classrooms can be a wonderful vehicle for learning some basic mathematical concepts. Children like to use the cubes for building all kinds of structures and teachers like to use them as concrete material for teaching addition and subtraction facts. But there is so much more that can be done with those beautiful cubes.

Using the cubes to cover plane regions

You can easily make many task cards like the ones in figure 1. Keep the dimensions of the plane regions in whole units so that the cubes fit "exactly." Some cards might be made for use with inch cubes, while others can be made for centimeter cubes. Remember that the cubes in your sets of Cuisenaire rods are centimeter cubes (1 cm × 1 cm × 1 cm).

Activities like these lead very nicely into counting by ones, twos, threes, and so on; linear measurement; and area measurement. And they can also be useful in developing or reinforcing basic concepts of multiplication and addition. You can encourage children to think of the cubes that cover a square region as "25 cubes" or "5 rows of 5 cubes" or "5 + 5 + 5 + 5 + 5." A rectangular region that contains 24 cubes might be thought of as "4 rows of 6 cubes" or "6 + 6 + 6 + 6" or "6 rows of 4 cubes." This provides a mental picture for multiplication as an array of cubes.

When he estimates the total number of cubes needed to cover the regions in (d), (e), and (f), the child has an application of addition. For example, he might think of (d) as having 4 cubes "one way" and 5 cubes "the other way," or 9 cubes altogether. The children will come up with a variety of ways to find the total number of cubes. That's great. In effect they are finding that a number has many names. Twelve can be 12, 3 + 3 + 3 + 3 or 4 + 4 + 4 or 10 + 2, and so on.

Many children may continue to count cubes by ones. Gradually they can be introduced to more efficient ways of finding

Photograph by Clif Freedman

How many cubes cover each of these? Guess first.

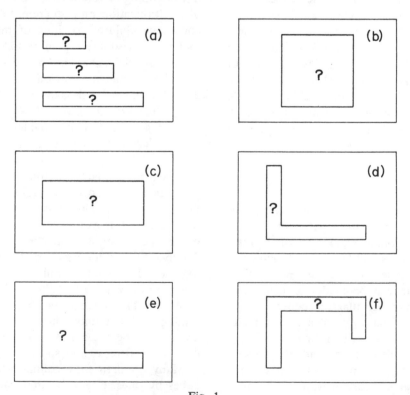

Fig. 1

Photograph by Clif Freedman

make the boxes out of oaktag. (What other patterns are possible?) You can use cellophane tape to hold the boxes together.

The children should get involved in making boxes. You can have children trace oaktag rectangular regions like the ones in

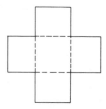

Cut along solid line and fold along dotted line. Then tape the corner edges together.

Fig. 3

the total number of cubes. You might, for example, make some color-coded cards like those in figure 2.

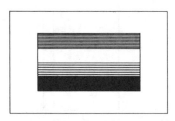

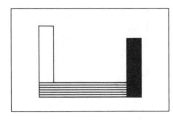

How many cubes cover each color?

How many cubes altogether?

Fig. 2

Filling oaktag boxes with cubes

Filling boxes with cubes can be a good introduction to the concept of volume or capacity. Homemade oaktag boxes are useful since they can be made so that the cubes fit "exactly" into them. A pattern like the one in figure 3 can be used to

figure 4 for the faces of the boxes. (How many different boxes can be made with a set of rectangular regions similar to these?) Also, have children transform the 3-dimensional boxes into the 2-dimensional patterns of the box by cutting along the vertical edges of the boxes. Then they should practice putting the boxes back together again. (Fig. 5.) This transformation can be very exciting for children.

The dimensions of the boxes should be in whole units: $2'' \times 2'' \times 2''$ or $4'' \times 2'' \times 2''$ or 10 cm $\times$ 5 cm $\times$ 5 cm, and so on. Actually, the dimensions of the boxes should be about $\frac{1}{8}$ of a unit larger than the intended measure so that the cubes can be put in and taken out more easily. A box with dimensions $2 \times 2 \times 3$, then, would really be $2\frac{1}{8} \times 2\frac{1}{8} \times 3\frac{1}{8}$.

Have a large assortment of boxes of various dimensions available. Give the children two or three boxes of different shapes (fig. 6) and have them guess how many cubes each box holds. Which box holds the most? Which box holds the least? Which boxes hold the same number of cubes?

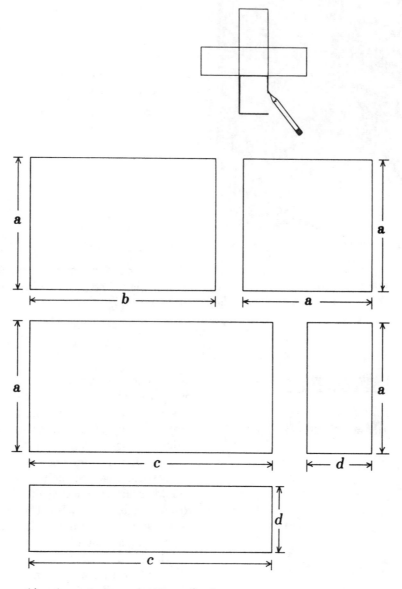

Use these to trace patterns for boxes.

Fig. 4

Multiplication develops naturally as the child attempts to find an easier way to calculate the total number of cubes in a box. The cubes covering the base of the box form an array, so multiplication can be used to find the total number of cubes there. Similarly, the cubes in the box can be seen as a certain number of "layers" of cubes (3 layers of 6 cubes, for example), which again involves multiplication.

Making boxes with squared materials

If the patterns for the boxes are drawn on squared paper, as in figure 7, you have an added advantage. You can introduce the concept of the surface area of the box. The child can calculate how many squares there are on each face of a box (each face is an example of a multiplication array) and how many squares there are altogether on the faces of a box.

You might also explore relationships between surface area and volume to show that they are distinct ideas. Does the box

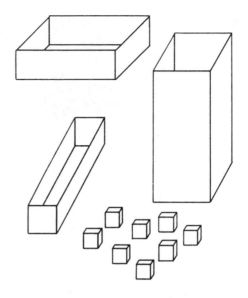

How many cubes does the box hold?

Fig. 6

Three — dimensional

cut

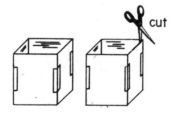

Open   Two — dimensional

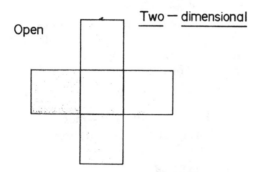

Three — dimensional

Put together again.

Fig. 5. Transforming a box

that has the most squares on its faces hold the most cubes? Can you find two boxes that hold the same number of cubes but have different numbers of squares on their faces? Can you find two boxes that

have the same number of squares on their faces but hold different numbers of cubes?

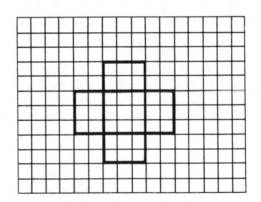

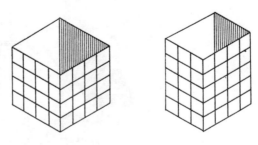

How many squares on the faces of boxes?

Fig. 7

By **James V. Bruni** and **Helene J. Silverman**
Herbert H. Lehman College, City University of New York

From blocks and model making to

The dynamic avenues for learning in the preschool and kindergarten can be vehicles for the learning of mathematics by older children. The background of experiences gained from these activities of early childhood, coupled with maturation, provides a base from which to explore new ideas with familiar materials. Block building and model making are two rich areas to consider.

BLOCK BUILDING

The many experiences with block building (using blocks of all sizes—modular blocks, floor blocks, and table blocks) give meaning to the language of comparison.

ratio and proportion

Children who use *big* and *little* as umbrella terminology for all comparative work can be helped to understand the subtle differences meant by *tall*, *wide*, *long*, *high*, *thick*, *short*, *narrow*, *thin*, and *low*.

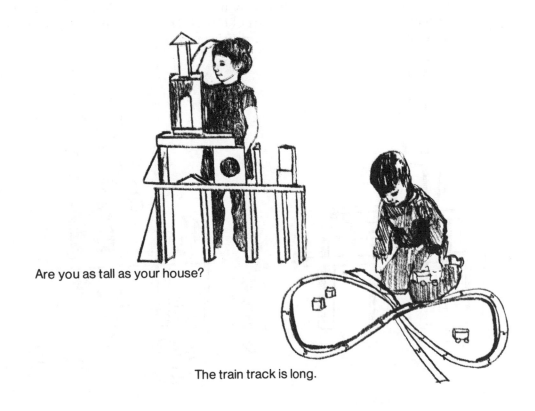

Are you as tall as your house?

The train track is long.

A teacher interacting with a young builder can help to provide meaning to such comparative vocabulary as *longer, wider, taller, twice as tall,* and *halving the length* as the child explores structures and compares the pieces in his set of blocks to each other and to his own body.

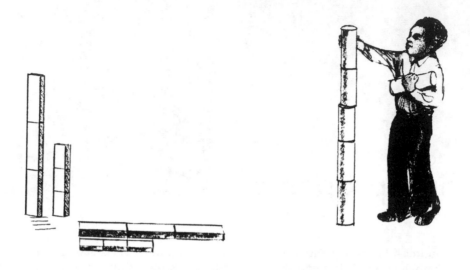

A game can be played by children who are experienced in block building. The leader builds a structure with a given set of blocks. The other players duplicate the "architecture" of the first player with a matching set of blocks.

Once children have facility in replicating structures, a "substitution" activity can be devised. In this activity, the figure must be copied, but players earn points if they can substitute other combinations of blocks to get the same structure. The fact that many pieces in a set of kindergarten blocks are multiples or divisions of the basic unit block makes it easy to find many variations in the combination of blocks.

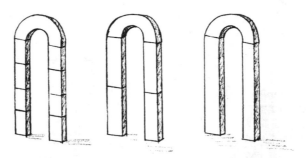

After experience with simple substitutions, a "one-difference" game may be introduced. In this game, the structure is changed in one aspect when it is replicated. For younger children, the change can be simply substitution of a block or series of blocks. For older children, the one change can mean a modification of one of the dimensions.

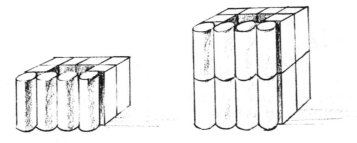

To focus on the changes in the dimensions, questions such as the following can be posed: How many blocks did you use for the first building? How many blocks did you use for the second building? How did the walls change? How did the door change? (If small, related cubes are available, the internal volumes of the structures may also be compared.) The game can be extended to include simultaneous change of more than one of the dimensions.

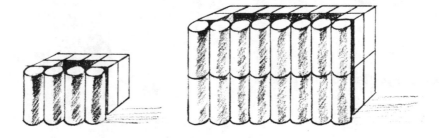

Different block sets with similar internal proportions can also be used. For example, a structure built from floor blocks can be copied with table blocks. The basic similarities between the two structures can easily be seen as the children learn to copy from larger or smaller figures.

USING ACCESSORIES

The addition of commercially produced or homemade accessories to the block-building activity can further encourage thinking about comparison and proportion as the children make judgments about the relative size of a hangar and an airplane, a garage and a car, and people and doorways.

Make a hangar for the airplane.

Make a garage. Can you find a car for it?

Build a house for these people.

GEOMETRY FOR GRADES K–6

Children who have had experience creating structures from "junk" materials can satisfy their desire to keep what they have made by reproducing improvised versions of their block structures. Paper towel tubes or coffee tins can be used in place of cylinders; cereal boxes, in place of blocks; and so on. You can assist the child with maintaining some of the basic proportions of the structure.

Setting a structure on brown paper permits the child to color in a setting for the building and provides additional opportunity for the child to consider the size of an object relative to the space surrounding it.

A supply of paper scraps and Play-Doh encourages the child to build activity into his creation and provides additional opportunities for the child to consider the proportional sizes of other objects in relationship to his structure. How wide must the door be for the children to get into the bus? How high must the door to the house be for all the people to go inside? Can you make people who can sit under the tree?

SCALE DRAWINGS

Junk structures also become cumbersome to keep. Scale drawings are a more feasible way of "saving" many structures, as well as an opportunity to relate physical objects to abstract symbols. Scale drawings also provide additional experience with the ideas of ratio and proportion.

Children learn to relate structures to drawings by matching blocks to a drawing. The various parts of the base of a structure can be outlined on brown paper and children could then be asked to reconstruct the building from the "blueprint" of the structure.

If a supply of small squares is available, the children can be taught to understand drawings in reduced size. You can mark off regions on the floor of the classroom with masking tape. The children then reproduce the marked-off space in miniature, using their small squares.

With experience, floor plans can be developed by using cardboard regions to represent the floor space taken up by objects in the classroom. The child counts the number of floor tiles covered by the objects in the room and then cuts out a region, in miniature, that will cover his squares in a like manner. Older children can be taught to shade regions on graph paper, eventually developing a floor plan for a classroom.

A "mystery object" game can be developed. Children would represent objects in the classroom by a scale drawing—this could be on graph paper. The scale drawings would be numbered and stored in a box. A child who had not been part of the drawing team would select a plan and then try to locate the object represented.

Children can develop a series of block structures and make scale drawings for each structure. A player can be given a set amount of time to match the drawings to the structures.

Other children may want to create "dream rooms" with matching scale drawings. The drawings would be given to other children who could then match the drawings to the actual rooms.

The school block can also be represented, at first by counting the number of sidewalk squares and reproducing them on graph paper. Buildings can also be represented by counting the number of sidewalk squares that border the building and then reproducing them accordingly on graph paper. At first the children may only be able to account for the width of the buildings, but with increased experience they can be taught to deal with the other dimensions simultaneously.

Children who have had experience measuring lengths with units can be taught how to represent the room to scale by measuring the room dimensions with metersticks and then representing the dimensions with toothpicks, Cuisenaire rods, or by drawing them on graph paper.

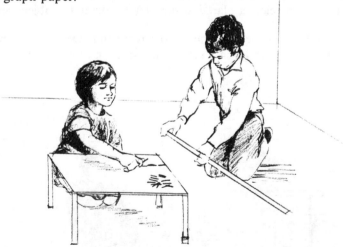

Activities like these provide experiences from which children can develop meaningful concepts of similarity, ratio, and proportion.

Patterns and Transformations

THIS group of articles focuses on the creation of patterns by doing something *to* or *with* simple geometric shapes. In some instances the patterns that evolve from the transformation of a figure are visually pleasing, even artistic. Sometimes, however, a transformation results in a distortion.

In "Living in a World of Transformations," Sanok identifies basic transformations and gives common examples of each. Morris ("Investigating Symmetry in the Primary Grades") describes a series of activities that develop appreciation of symmetry, an element of both patterns and transformations. The next two articles describe activities in which children generate designs or patterns with a square. In "Mathematics and Art from One Shape," Ibe has children trace a square to create patterns. In "Making Patterns with a Square," Bruni and Silverman use thirteen basic patterns, colored regions within squares, to make and analyze other patterns.

Van de Walle and Thompson ("Concepts, Art, and Fun from Simple Tiling Patterns") focus on tessellations (tiling patterns), extending the notion of patterns still further. In "Pictures, Graphs, and Transformations: A Distorted View of Plane Figures for Middle Grades," Swadener introduces the notion of transformations on a plane in a way that children can comprehend. Some of the results pictured in the article look strange, but they have real-world applications. Teachers and students may gain a better appreciation of problems in map making and the distorted appearance of Greenland, for example, on a Mercator projection.

LIVING IN A WORLD OF TRANSFORMATIONS

By **Gloria Sanok**

We live in a world of transformations. Some transformations change the size or shape of an object or figure (Swadener 1974). Some transformations change neither. Transformations that change neither the size nor shape are *translations* (slides), *rotations* (turns), and *reflections in a line* (flip). A *dilation* is a transformation that changes the size of an object, but does not change its shape. *Congruency* and *similarity*, which are often studied initially in the context of triangles, may be learned more intuitively through transformations.

Transformations

A *translation* of an object or figure may be thought of as a slide along a straight line. A translation has both direction and distance. Thus in figure 1, the nose and the toes of the child on the slide have moved the same distance and since they have moved along parallel lines, they have also moved in the same direction.

A *rotation* of an object in a plane is a turn about a point. Every rotation has a center and an amount of turn. The child in figure 2 has rotated the flag one

fourth of a turn (90°); his hand is the center of rotation.

A *reflection* of a plane figure is obtained if the figure is flipped across a line. The reflection reminds us of a mirror image. The *line of reflection* is the perpendicular bisector of every line segment that joins a point in the figure and its image in the reflection. Thus, the line segments joining the corresponding points of the *F*s and *N*s in figure 3 are bisected by the line of reflection.

A *dilation* (enlargement) of a figure is obtained by multiplying each of the

Fig. 1

Fig. 2

Fig. 3

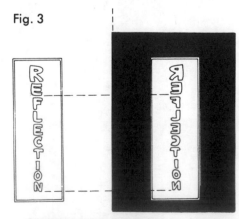

GEOMETRY FOR GRADES K–6

distances between a fixed point (C in figure 4) and a point of the figure by the same number. Thus the distance from C to E_2 is twice the distance from C to E_1; the distance from C to T_2 is twice the distance from C to T_1. The distance between C and every point of the large cat is twice the distance between C and the corresponding points of the small cat.

Figures that have the same size and the same shape are said to be *congruent* to each other (fig. 5). Figures that have the same shape but not necessarily the same size are *similar* to each other (fig. 6). All congruent figures are also similar. One or more of the *transformations, translation, rotation,* and *reflection,* may be used to "move" one congruent figure onto another. A *dilation* and one or more of the *transformations, translation, rotation,* and *reflection,* may be used to "move" one similar figure onto another.

Symmetry

Symmetry is useful in designs and evident in nature, art, and architecture. Designs in nature illustrate a natural tendency to make things balance and we look on symmetry as a type of balance.

An object has *line symmetry* if it *matches itself* when reflected in a line. The uppercase letter A, for example, is symmetric with respect to a line (fig. 7).

Fig. 7

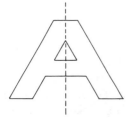

An object has *rotational symmetry* if it *matches itself* when rotated about a point. The carousel in figure 8 matches itself when rotated one third of a turn about its center. An object has *point symmetry* if it *matches itself* when rotated a half turn (180°) about a point.

Gloria Sanok is a teacher and inservice instructor in the Wayne (New Jersey) public schools. She is also an adjunct instructor at William Paterson College.

Fig. 4

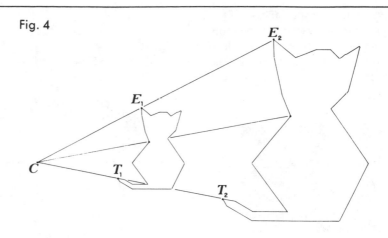

Fig. 5

Fig. 6

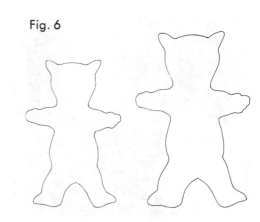

Fig. 8

Point symmetry is a special case of rotational symmetry (fig. 9). The monkey figures in figure 9 are symmetric about point *P*. The bracelet in fig. 9 is another example of point symmetry.

Fig. 9

a.

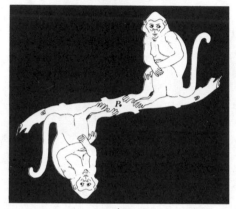

b.

Numerous examples from the "real" world can be provided for the study of transformations and symmetry.

Examples of Transformations in Nature

An apple, if cut in an unconventional way, illustrates rotational symmetry—a beautiful five-pointed star (fig. 10). If the star is rotated about the center one fifth of a turn, it matches itself. An orange can be cut in half horizontally to show rotational symmetry and vertically to show line symmetry. A snowflake pattern has three lines of symmetry (fig. 11). It also has rotational symmetry and point symmetry.

A butterfly is a beautiful example of an object that is symmetric with respect to a line (fig. 12). Leaves are also excellent examples of objects that are symmetric with respect to a line (fig. 13).

A honeycomb illustrates translations, rotations, and reflections (fig. 14). The starfish and many other seashells show rotational symmetry. Many flowers also illustrate rotational symmetry.

Fig. 10

Fig. 11

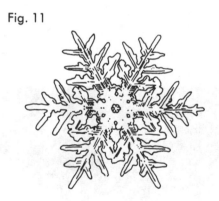

Fig. 12

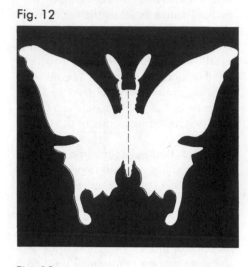

Fig. 13

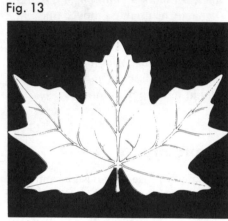

Fig. 14

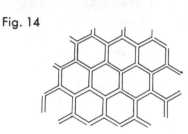

Other Examples of Transformations around Us

Transformations can also be found in the repeated patterns of many gift-wrapping papers and wallpapers (fig. 15). Wheels and propellers show rotational symmetry. Dinner plates with specific markings provide excellent patterns for discussion (fig. 16). In art, the graphics of M. C. Escher are perfect examples of transformations (fig. 17).

The idea of congruence is used in everyday life. When we sort out coins from a handful of change, we use congruence. Coin-sorting and vending machines use the congruence of coins of

Fig. 15

Fig. 16

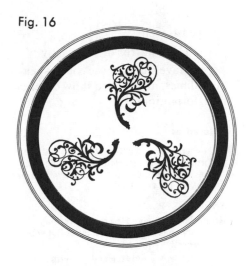

Fig. 17

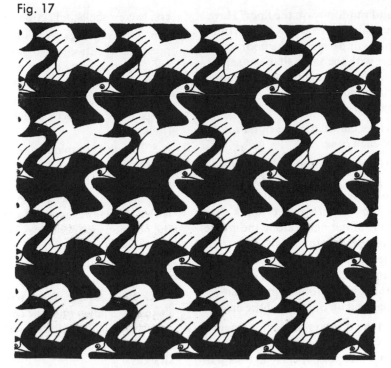

M. C. Escher—Study of Regular Division of the Plane with Birds, 1955.
Escher Foundation—Haags Gemeentemuseum—The Hague.

the same denomination to distinguish between the coins.

A pie (or cake) can be divided into congruent pieces. Each piece (sector) represents a rotation from one piece to the other about the center of the pie.

Examples of translations, reflections, rotations, and dilations are all around us. We live in a world of transformations.

Teaching Aids

A "shooting gallery" provides an excellent example of translations, rotations, and reflections. The materials needed to make a "shooting gallery" include an ordinary piece of 56-by-70-cm black oaktag, a paper plate, a

tongue depressor, five yellow cutout ducks, a paper fastener, and some tape (fig. 18).

The letters of the alphabet illustrate vertical line symmetry, horizontal line symmetry, point symmetry, or none of these. Make a set of 10-cm high letters of the alphabet so that the letters can be folded vertically, horizontally, or

both. Attach paper fasteners to the letters that have point symmetry in order to show the half turn (180°) (fig. 19).

Activities

1. Take a variety of cutout shapes and designs, including duplicates, and toss them up into the air so that they

Fig. 18

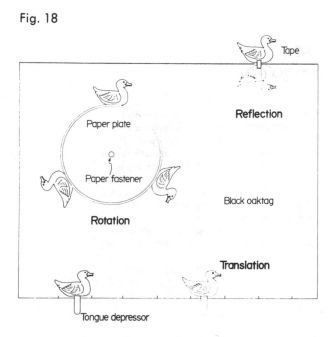

Fig. 19

A B C D E F
G H I J K L
M N O P Q R
S T U V W X
Y Z

will land randomly on the floor. Have the students describe how to get from one shape or design on the floor to another with the same shape or same design. Did the student have to translate (slide), rotate (turn), or reflect (flip) the shape or design?

2. Have children act as mirror images of each other.

3. Have students make designs with folded paper. Each student folds a piece of paper, then opens it, and using a wide-nib pen with any color ink, writes his or her name along the crease in the paper. Before the ink is dry, they fold the paper again. What does the result indicate? Students can do the same thing with drops of ink or patterns painted on one side of the crease.

4. Make displays of student fingerprints. (Make a large smudge of pencil lead on a piece of paper. Rub the tip of one finger on the smudge. Put the sticky side of a piece of clear cellophane tape over the smudged fingertip. Lift the fingerprint off onto the tape. Put the tape with the fingerprint on paper for display.) Have students compare their fingerprints.

5. Take photographs of things in nature that show transformations—flowers, leaves, shells, reflections in a lake.

6. Collect leaves. Wax the leaves or press leaves between two pieces of waxed paper, using a very warm iron. Or make blueprints of leaves.

7. Make plaster casts of leaves, twigs, flowers, insects, or even animal tracks.

8. Using a compass, make a variety of colorful circular designs. Display the designs on the bulletin board.

9. Make a "snowflake" by starting with a regular hexagonal polygon. Fold the polygon along the lines of symmetry, then cut out a design. Unfold to show the snowflake.

10. When it snows, place a piece of black cloth outside. When a few snowflakes have fallen onto the cloth, look at the snowflakes through a magnifying glass. Examine the symmetry in their designs.

11. Have students find samples of gift-wrapping paper or wallpaper patterns and identify the kinds of transformations that are used in the patterns.

12. Have students bake cutout cookies of the three bears or gingerbread boys of different sizes. The cookies provide an excellent lesson on enlargements (similarity) and can be enjoyed as well.

13. Use a mirror or Mira activity.

14. Have students use coins—pennies, dimes, nickles—to show different transformations.

References

Swadener, Marc. "Pictures, Graphs, and Transformations: A Distorted View of Plane Figures for Middle Grades." *Arithmetic Teacher* 21 (May 1974):383–89.

See also:

Bruni, James V., and Helene J. Silverman. "From Shadows to Mathematics." *Arithmetic Teacher* 23 (April 1976):232–39.

Bruni, James V., and Helene J. Silverman. "Making Patterns with a Square." *Arithmetic Teacher* 24 (April 1977):265–72.

Johnson, Martin L. "Generating Patterns from Transformations." *Arithmetic Teacher* 24 (March 1977):191–95.

Kidder, F. Richard. "Euclidean Transformations: Elementary School Spaceometry." *Arithmetic Teacher* 24 (March 1977):201–7.

Krause, Marina C. "Wind Rose, the Beautiful Circle." *Arithmetic Teacher* 20 (May 1973):375–79.

Morris, Janet P. "Investigating Symmetry in the Primary Grades." *Arithmetic Teacher* 24 (March 1977):181–86.

Zweng, Marilyn. "A Geometry Course for Elementary Teachers." *Arithmetic Teacher* 20 (October 1973):457–67.

A bibliography, "Transformation Geometry Information Resources," is available from NCTM Headquarters office.□

Readers' Dialogue

The April 1978 article "Living in a World of Transformations," equated point symmetry to 180° rotational symmetry for any object.

Point symmetry (or center of symmetry) exists when an object maps onto itself after every point has been translated in a straight line through the center. The tetrahedron is a good example of an object with point symmetry but which does *not* have 180° rotational symmetry.

Only for the limiting case of two dimensions can we always equate point and 180° rotational symmetry.

Jack O'Neill
St. Jerome's School
Hamilton, Ontario

Investigating symmetry in the primary grades

JANET P. MORRIS

An assistant professor of mathematics education at the University of Michigan-Flint, Janet Morris teaches mathematics content and methods courses for elementary and secondary teachers.

The children were thoroughly engrossed in investigating the symmetry of the letter *S*.

"I think it folds."

"No, it can't, because if you folded it here, these two parts wouldn't match. They'd be going different ways. Let's try turning it."

"Yes, that will work! See, it takes two part turns." These were second-grade children and what they were doing was part of a week-long unit that integrated the concepts of bilateral and rotational symmetry.

Symmetry was the chosen topic for several reasons. First, it is a familiar idea. Objects that are well-balanced, well-proportioned, and have regularity of form abound in the environment. Many works of nature—leaves, snowflakes, even humans themselves—exemplify symmetry. Recognizing the aesthetic appeal of things that are duly proportioned, man has created works of art and architecture that are symmetric. This overt presence of symmetry in the environment makes it of natural interest to children.

Secondly, symmetry lends itself to intuitive interpretations that can be investigated through informal, hands-on manipulations. Such activities seem to be more closely related to the way the young child thinks and learns than some more formal geometric experiences that rely heavily on vocabulary and abstract perceptions.

Thirdly, symmetry is good mathematics, and early work with the subject can provide foundations for extensions in several directions.

As a result of considerations such as these, bilateral symmetry is appearing in many current elementary texts. A design has bilateral symmetry with respect to a line *l* if for every point *P* on the design, there is a corresponding point P′ the same distance from *l* but on the opposite side. (Fig. 1a) This concept has an intuitive but mathematically sound interpretation that can be physically tested by a young learner. By folding the design along the line of symmetry, it can be seen that the parts of the design "match."

Similarly, a design has rotational symmetry about a point *O* if every point *P* of the design can be made to correspond with another design point P′ when the design is rotated about *O* by less than a full turn. (Fig. 1b) This type of symmetry also has a test that young learners can physically apply. By tracing the design, then turning the tracing part of a full turn, they can readily determine if the design can be made to match itself.

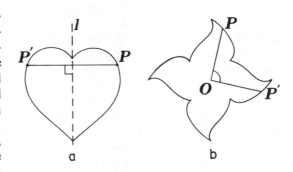

Fig. 1

Although rotational symmetry is not included as frequently as bilateral symmetry in primary texts, the results of this unit indicate that the concept is at least as easy for children to understand, and there are particular benefits when the two ideas are taught together. Awareness of both types provides the students with an alternative explanation for the balance of a design, when the test for one type of symmetry fails, as indicated in the children's dialogue. If only bilateral symmetry is taught, students are puzzled by figures, such as the S-shape and the parallelogram, which have like or congruent parts but fail the fold-and-match test. Also, presenting both types together and using tests for each that involve motions provide a foundation for further work in motion geometry.

The first lesson of this primary level unit introduced designs that could either be matched to themselves by folding on a single line or by turning a tracing of the design. Later lessons dealt with designs that could be matched both ways. Using an overhead projector, the teacher asked the children first to consider a butterfly design. They quickly pointed out that parts of the design were just alike—the two wings, the two antennae, and so forth. A cutout of half the design brought out the fact that the matching of like parts could not be achieved by sliding one half onto the other; rather, folding along a line through the middle caused the parts to "match exactly." Next a block letter N was considered. Experimenting revealed that although parts of this design were also just alike, folding would not match the parts. After the children tried several different ways of folding it, it was suggested that this design had a different kind of matching. At the teacher's suggestion, they then traced it, and turned the tracing part of a full turn to find it matched the original design that had been held still. By having the children leave the original design on the table and move just the tracing, difficulties were avoided that might result from moving both the original and the tracing, especially when such a composition results in the identity.

A mark at the top of the tracing helped the children remember its original position. The butterfly design was investigated with a tracing, and the children concluded that it matched with a fold but not a turn. After the class investigated an asymmetrical design that could not be made to match itself with either a fold or a turn, the children were given individual packets containing designs that they then tested for folding and turning matchings. (Fig. 2) The designs for this first lesson were chosen to have either one line of symmetry, or turning symmetry, or neither.

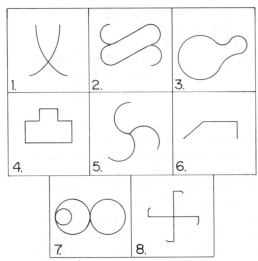

Which designs match themselves by folding along a line? Which match by turning a tracing part of a full turn? Do any not match either way?

Fig. 2

These packets, containing designs appropriate for the particular lesson, were used throughout the unit to encourage individual and small-group investigations and explorations. Providing the student with his own set of designs allowed him to make his own prediction of the result of a particular manipulation, and then determine for himself the accuracy of his prediction. Such procedures are especially important for the development of perceptual ideas at this age level.

Although the designs used were two-dimensional ones, which could be precisely constructed to have the required symmetry, transfer of the concept to real-world objects was natural for the children. With only the suggestion that they look around themselves, they were volunteering that wheels, windows, fire hydrants, even their own bodies exemplify symmetry. Investigating these familiar objects for symmetry emphasized the relation of geometry to the child's world.

In the next lesson, the children noted that when designs were folded so that the parts matched, the fold was along a line. The term *line symmetry* was introduced to refer to this property, and was readily used by the children. The notion of line symmetry as a reflection about the line was then brought out by having the children place the edge of a mirror on the line of symmetry of a design and perpendicular to the design, and then see that they could see the whole design, half of it reflected in the mirror, the other half on the paper. Each child was given a nonbreakable army field mirror, obtained at the local surplus store, to investigate the designs in his packet. (Fig. 3) Several designs had more than one line of symmetry. Designs with two lines of symmetry were easy for the children, perhaps because the lines were mutually perpendicular. Designs with more than two

lines were more difficult, but were usually correctly found using the fold test, then checked with the mirror. The circle was particularly fascinating, and after much thought, it was concluded that the circle has "lots" of lines of symmetry.

Using the mirrors as a check for a line of symmetry first found by folding provided a meaningful sequence for the mirror work. The results of placing the mirrors in various positions were especially fascinating to the children. Soon they were out of their packets—investigating the letters of their names, pictures in books, objects on the bulletin board, anything and everything. The number of independent investigations that occurred spontaneously among the children with the mirrors supports suggestions that units devoted to mirror investigations are both suitable and valuable (Abbott 1970 and Walter 1966).

The third day's activities involved rotating cutouts of familiar plane figures to see if they could be matched to their tracings with part of a full turn. (Fig. 4) To call this

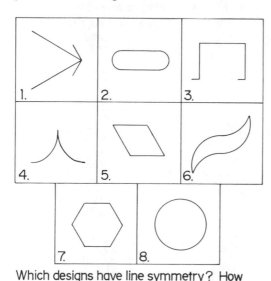

Which designs have line symmetry? How many lines of symmetry do they have?

Fig. 3

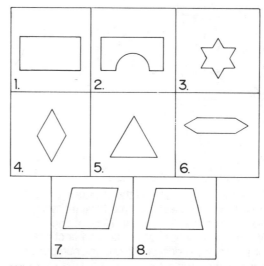

Which shapes can be turned part way around and look just as they did before? Which can be folded so their parts match? Do some do both? What is alike about shapes 1 and 2? What is different? What do you notice about shapes 7 and 8?

Fig. 4

property "turning symmetry" seemed reasonable to the children—the figure was turned to match itself, and symmetry indicated a matching, just as it had in line symmetry.

Most of the children needed no prodding to notice that some of the figures, such as the rectangle, not only had turning symmetry, but also line symmetry. The children were not confused by the presence of both types in a single figure. It was an easy task for them to classify a figure as having both types, as having one without the other, or as having neither. Further, certain relationships between the types were recognized, making this integrated approach even more attractive. One such relationship, that a design with at least two lines of symmetry also has turning symmetry, was brought out in the investigations of the fourth lesson. (Fig. 5) The children then used this idea to make

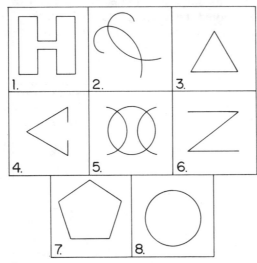

For each design, find all the lines of symmetry. Now, before you test it, do you think it has turning symmetry? Test it. Were you right? What do you notice about the number of lines of symmetry a design has and the number of part turns it makes?

Fig. 5

their own "snowflake" designs by folding a square of paper into quarters and cutting along the nonfold edges. (Fig. 6) They were delighted with their results, and eagerly

checked each other's designs for both types of symmetry.

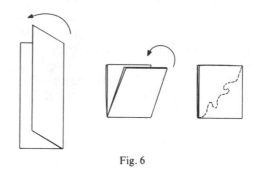

Fig. 6

For those designs that had turning symmetry, the children indicated how many "part turns" it took to return the design to its original position. Thus they said the equilateral triangle takes "three part turns." Later work, discussing degrees of rotation, could make this notion more precise.

Another idea that could be developed in later work is that of center of rotation—that there is a unique point about which the design is turned. This idea could be extended further by showing that a design with turning symmetry can be generated by part of the design being turned about the center (Fig. 7), in a way similar to a design with line symmetry being generated by half the design being flipped across the line of symmetry.

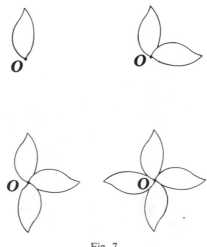

Fig. 7

Two worksheets were included in the last lesson of the unit. The first (Fig. 8a) involved identifying which jigsaw puzzle piece fit in a given hole by mentally rotating the given pieces to see which would fit. Most of the children found the task easy, and said they could check their answer by showing that a tracing of the piece had the same shape as the hole.

The second worksheet (Fig. 8b) was half a design on a grid, which the children completed so as to have line symmetry. As they worked, they continually checked their additions to the design in two ways, by folding to make sure the parts matched the given parts, and by counting the spaces of the grid to see that the added parts were the same distance from the line of symmetry. The latter task was more difficult, especially drawing corresponding horizontal lines. Some children had a tendency to draw the image line going in the same, rather than opposite, direction. However, the fold test quickly corrected this. The finished design had two lines of symmetry, and so also had turning symmetry. The children then colored their designs, matching the colors of corresponding parts to keep the symmetry of the design.

One of the valuable strategies, used to complete the grid design and emphasized throughout the unit, was the idea of predicting an outcome, then checking the accuracy of the prediction by performing the task. Teachers will recognize the value of this predict-then-check technique in other areas of mathematics.

This second-grade unit on line and turning symmetry barely skimmed the possibilities. With a small amount of class time, the rewards were great. The children's enthusiasm and interest alone might have been sufficient justification for the unit, but in addition to that, the children learned more geometry in general by finding more properties of familiar figures, such as rectangles and triangles, and of new designs. Although not explicitly mentioned, their work with symmetry involved the concept of congruence. It also involved visualizing the motions of flips and turns on designs,

and realizing that size and shape are invariant under these motions. In all the activities, the children were becoming more aware of the shapes around them, and they were learning to see the balance of many

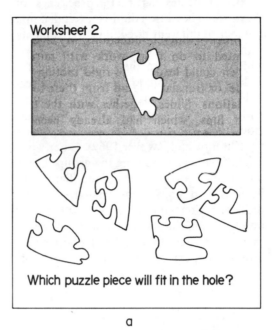

Fig. 8

things as examples of symmetry. The activities thus fit the needs and interests of young children as they interpret their real-world space experiences.

Activities like these could be adapted to a range of grade levels by varying the complexity of the designs and the preciseness of the work. They also could be extended in a number of different directions. My students wanted to do more work with mirrors, which could lead to the rigid motion of a slide, or translation, and from there to tessellations. Slides, together with the turns and flips, which had already been introduced, lead to topics of motion geometry (Phillips and Zwoyer 1969). Later work might involve considering the groups based on symmetries of such plane figures as squares and equilateral triangles. In all these directions, the students are informally investigating shapes and finding more of their properties. The study of symmetry is thus made part of a continuous thread of geometrical experiences, not an isolated activity.

In summary, symmetry was shown to be an appropriate topic for primary-grade work. It can be informally developed through physically manipulating objects, and thus, is pedagogically sound. It is of high interest to children because of its close relation to their environment. And it is a mathematically sound basis for later work in such areas as congruence, groups, and motion geometry.

References

Abbott, J. S. *Mirror Magic.* Chicago: Franklin Publication, 1970.

Phillips, J. M., and R. E. Zwoyer, *Motion Geometry Books 1, 2, 3 and 4.* (Based on earlier editions by UICSM staff members) New York: Harper and Row Co., 1969.

Walter, Marion. "An Example of Informal Geometry: Mirror Cards." *Arithmetic Teacher* 13 (October 1966):448–52.

Walter, Marion. "Two Samples of Informal Geometry for Young Children." Ph.D. dissertation. Harvard University, 1967.

Mathematics and art from one shape

MILAGROS D. IBE

Ontario Institute for Studies in Education
Toronto, Ontario, Canada

One of the sad things about many current lessons on plane figures in elementary school geometry is the teaching of shapes only in terms of what they appear to be and not in terms of their potential or hidden possibilties for mathematical understanding and beauty.

Take the square, for example. Teachers belabor its properties of equal-sidedness, its four right angles, and possibly its four axes of symmetry. Beyond these, few other facts are discussed about it; but it is because of these very properties that many special forms can be produced from this shape.

Consider one square-shaped board or tile that can be traced by a child. A teacher may ask how the shapes shown in figure 1 can be drawn from the square.

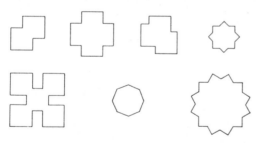

Fig. 1

At the start it may be hard to get pupils to see how, because perceptual closure is involved in these figures. Hence, the teacher may need to draw the complete squares, as shown in figure 2.

It is the properties of equal-sidedness and symmetry that make it possible to get **b, d, e, f,** and **g.** Children can then be made

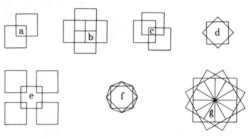

Fig. 2

to appreciate these properties more by getting them to try the same methods with a rectangle or a trapezoid. They will not find so many exciting figures formed as with the square.

It is quite likely that pupils will be fascinated by **d, e, f,** and **g** in figure 2 and will be motivated to produce similar ones or one like that shown in figure 3.

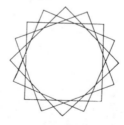

Fig. 3

From this shape (fig. 3), the child can infer that the more he spins and traces the square about a point, the more closely he defines a circle inside and that connecting the outer points of the traces will also define a circle. From these observations he will be close to intuiting what would otherwise be too abstract for him to comprehend—the fact that a circle is a polygon of *infinitely* many sides.

The shapes produced can be the basis of arithmetic problems. For example, if the area of the basic square is known (say, 12), the pupils can be asked to estimate the areas of **a, b, c, d,** and **e** in figure 2. The possibilities are limitless. For geometry practice, the pupils can be told to identify and count the shapes of each kind within the new shapes in figure 2.

The same approach can be used with other basic shapes. Consider, for example, the shapes shown in figure 4, which can be traced and used to generate others. Some of the possibilities are given in figure 5.

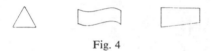

Fig. 4

Many basic shapes come in plastic, see-through templates that young children will enjoy tracing. These templates are very handy, especially when children are not yet skilled in using rulers or straightedges. Poorly developed motor skills for tracing straightedges and cut patterns need not be a hindrance to a teacher's presenting exercises like the above if templates are used.

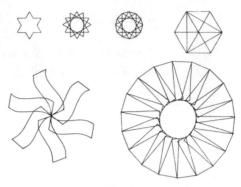

Fig. 5

To summarize—triangles, squares, rectangles, trapezoids, and other common shapes will not be dull learning material if pupils can be taught to develop insights as to possibilities these shapes have beyond their basic mathematical significance. Teachers need only to draw a pupil's attention and interest to what can be done with one shape for a start. From there, the pupil can come up with his own creations using other basic shapes. The new figures formed can then be discussed for both their artistic and mathematical implications.

By **James V. Bruni** *and* **Helene J. Silverman,**
Herbert H. Lehman College, City University of New York

Making patterns with a square

In a previous "Let's Do It!" (*Arithmetic Teacher,* February 1977) ways to use a square to help develop fraction concepts were suggested. This month we will take another look at that square. In fact, we will do more than look at it. We will "slide it," "turn it," and "flip it," and create all sorts of geometric patterns. Beginning with a square partitioned into eight (then later, sixteen) congruent triangles, the activities that follow suggest ways for children to develop some basic ideas about congruence, symmetry, and geometric transformations (translation, rotation, reflection) while they enjoy many mathematical and artistic experiences and develop their problem-solving abilities.

DESIGNS ON A SQUARE

Starting with a sheet of squares, each partitioned into eight congruent triangles, tell children to choose a colored crayon and then color the interior of each square so that in each square half of the triangular regions are colored and half are not (fig. 1). Encourage the children to try to make as many *different* designs as possible. Sometimes it is easy to determine that two designs are just the same, but for other pairs of designs it is more difficult to judge (fig. 2).

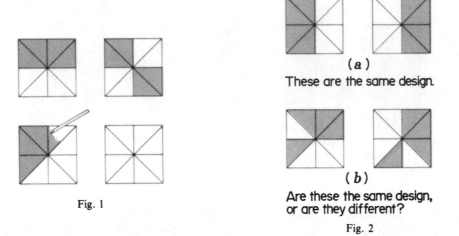

(a)

These are the same design.

(b)

Are these the same design,
or are they different?

Fig. 2

Fig. 1

How can a child determine whether a design is the "same as" or "different from" another design? It is useful to have the children suggest ways to prove that two designs that look different are really the same (or actually different). One way is shown in figure 3.

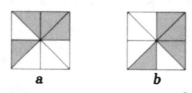

a   b

Is design *a* the same as design *b*?

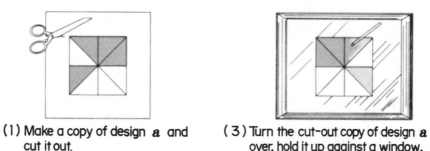

(1) Make a copy of design *a* and cut it out.

(3) Turn the cut-out copy of design *a* over, hold it up against a window, and color the back where the front has been colored.

Flip   Turn

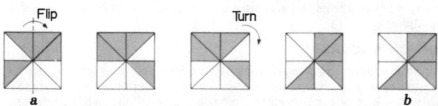

a   b

(3) Flip your cut-out of design *a* and/or rotate it in different ways and try to make design *a* look just like design *b*.

Fig. 3

Through activities like these the children begin to realize that although two designs may seem like different designs, if one design of a pair can be flipped and/or rotated so that it looks just like the other design, then the two designs are *not* different. With that understanding the children can proceed to find as many *different* designs as possible in which half of the triangular regions inside each square are colored. How many *different* designs are there? 8? 12? An unlimited number?

Finding how many designs there are is an excellent problem for individual children or small groups of children to try to solve. How will they know when they have the maximum number of different designs? How can they "prove" that they have found the maximum number and that no one can find another different design? Questions like these can lead to some lively discussion and interesting problem-solving strategies, and the children may be able to discover that exactly thirteen *different* designs are possible (fig. 4). Any design made by coloring half of the triangular regions can be shown to be the same as one of these thirteen basic designs.

You can develop similar activities using a square whose interior has been partitioned into sixteen congruent triangles (fig. 5). How many different ways are there to color the inside of that square so that one half of the triangular regions is in color? Figure 6 offers some possibilities. Again the question of how to determine whether designs are the same or different is important, and again, by making a paper model of one of a pair of designs and rotating or flipping it, it is possible to show whether the designs can look alike.

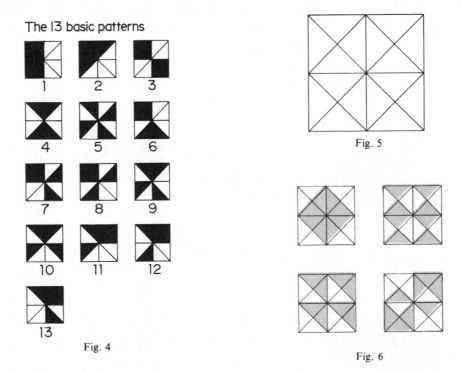

The 13 basic patterns

Fig. 4

Fig. 5

Fig. 6

You can extend these experiences further by introducing other directions for completing the designs. For example, figure 7 has been colored according to the following direction: one-fourth one color, one-half another color, one-fourth no color (or a third color). How many different designs can be made using this rule? Or, similarly, how many designs could you make that are one-eighth one color, five-eighths another color, one-fourth no color (or a third color)? Other similar directions can be used—the children can make up their own rules—and other types of designs will result. You may wish to organize bulletin board displays of the many designs developed by using different rules.

$\frac{1}{4}$ one color

$\frac{1}{2}$ another color

$\frac{1}{4}$ no color (or a third color)

Fig. 7

REPEATING A FAVORITE DESIGN

As the children develop different designs, you can ask them which designs are their favorites and why. You might collect the data and show it in graph form, as in figure 8. Very often children prefer more symmetrical designs even when they have not

Which is your favorite design?

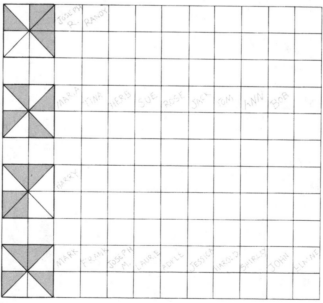

Fig. 8

been formally introduced to ideas about symmetry. A discussion of favorite designs is a good opportunity to talk about special properties of some designs—bilateral and rotational symmetry, for example—as suggested in figure 9. After the children have chosen their favorite designs, you can encourage them to develop some interesting patterns using those designs. Figures 10 and 11 indicate some ways of doing this.

GEOMETRY FOR GRADES K–6

(1) This design can be flipped and looks just the same.

(2) This design can be rotated and looks just the same.

(3) This design can be flipped or rotated and looks just the same.

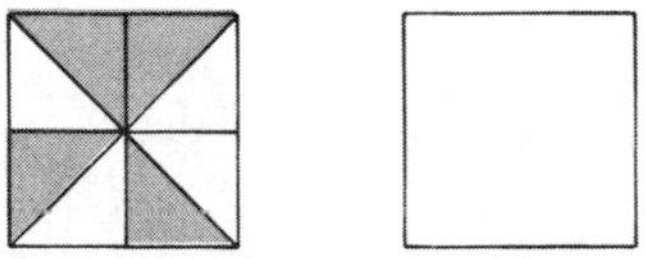

(4) Can you rotate or flip this design and have it look the same?

Fig. 9

Make four copies of this design, then cut out and color the back of each design.

Place the four copies in a row in different ways to make different patterns. How many different patterns can you make?

Can you make these patterns?

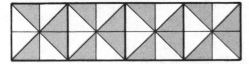 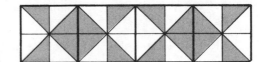

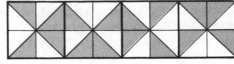

Fig. 10

Sometimes you want to describe a pattern so that you can make it again. Here is one way to do this. First, show which motions you are going to use and how you are going to indicate the motions.

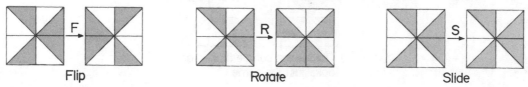

Flip   Rotate   Slide

Then write a rule to describe a pattern. Here are two examples.

Rule   Pattern

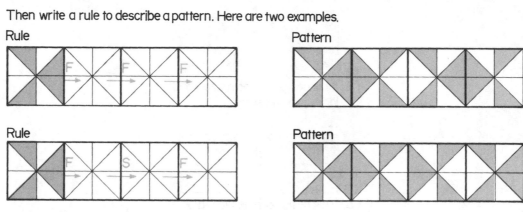

Rule   Pattern

Can you make the pattern that goes with the rule below? Can you continue the pattern to the right or to the left?

Rule   Pattern

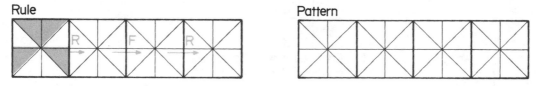

Make up some patterns that you like and then write the rule for your pattern. Ask someone to try to make the pattern from your rule.

Fig. 11

Patterns in two directions

Once the children are able to create and reproduce patterns along a horizontal line, they can be introduced to making designs with both horizontal and vertical motions (fig. 12).

Rather than color each square individually, the children can make design "stamps," similar to rubber stamps, and create stamped designs with paint or ink. The square designs can be cut from foam pads with self-adhesive backing, which are available commercially. The final design can be attached to a square block of wood, which then can be used as the "rubber stamp" (fig. 13). If a class wants a project, a choice of one or more of the favorite designs can form the basis of some creative patchwork quilts.

Can you follow the rules below to make patterns?

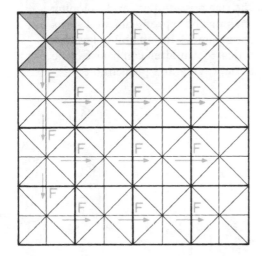 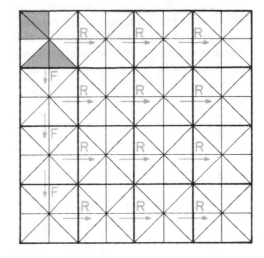

Can you write a rule for getting the pattern below.

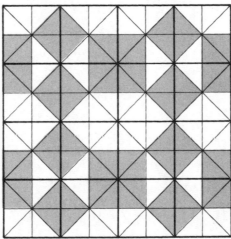

Fig. 12

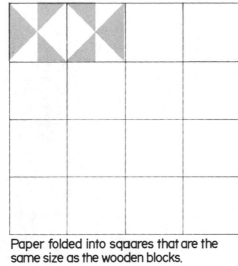

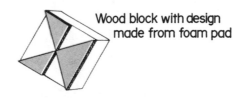
Wood block with design made from foam pad

Paper folded into squares that are the same size as the wooden blocks.

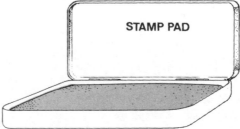
STAMP PAD

Fig. 13

The preceding activities only begin to suggest the rich possibilities for developing worthwhile mathematical activities involving the making and reproducing of designs and patterns. Pattern activities provide learning opportunities for all ability levels—they can be made very simple or surprisingly sophisticated. Children can have a basic introduction to transformations of shapes that is challenging and pleasurable.

Let's Do It

Concepts, Art, and Fun from Simple Tiling Patterns

By **John Van de Walle**, *Virginia Commonwealth University, Richmond, Virginia* *and* **Charles S. Thompson**, *University of Louisville, Louisville, Kentucky*

A tiling pattern (or tessellation) is a repetitive pattern, something like a mosaic, using one or more geometric shapes. The tiles in a tiling pattern should fit together so there are no holes or gaps between them. They are placed in a repeated pattern so that once begun, a huge plane surface could be covered by simply repeating the pattern over and over.

Tessellations can become quite complex, as the works of M. C. Escher will attest. In recent years tessellations have become popular as enrichment activities for junior and senior high school students. Even kindergarten children can create beautiful tiling patterns if the subject is approached correctly. Moreover, some very nice mathematics can grow out of discussions of the patterns once they have been made.

The Basic Tile Shapes

A little experimentation quickly leads to the conclusion that some shapes are easier to tile with than others. We suggest the four shapes drawn in figure 1 as good beginning tiles. Although many other shapes could be used, these four will provide some variance in level of difficulty and an abundance of design possibilities.

Because children need to do a lot of experimenting with their designs, we have found tiles cut from corrugated cardboard work very well at the beginning stages. Collect a lot of corrugated cardboard from boxes. After the boxes are cut into flat pieces with a mat knife, a paper cutter can be used to cut large numbers of each tile. To cut the diamonds and triangles, first measure and cut strips as shown in figure 2. Mark these using a pattern guide and then cut the individual tiles. Two children working together will need thirty to forty tiles. Once the tiles are made they can be used for several years.

What Is a Pattern?

To explain what a tiling pattern is will be one of the most difficult tasks of this project, especially with young children. The best way is to find some familiar examples. Perhaps the most common are the brick patterns in a wall, a checkerboard, a honeycomb, floor tiles, ceiling tiles, and brick sidewalks. If the

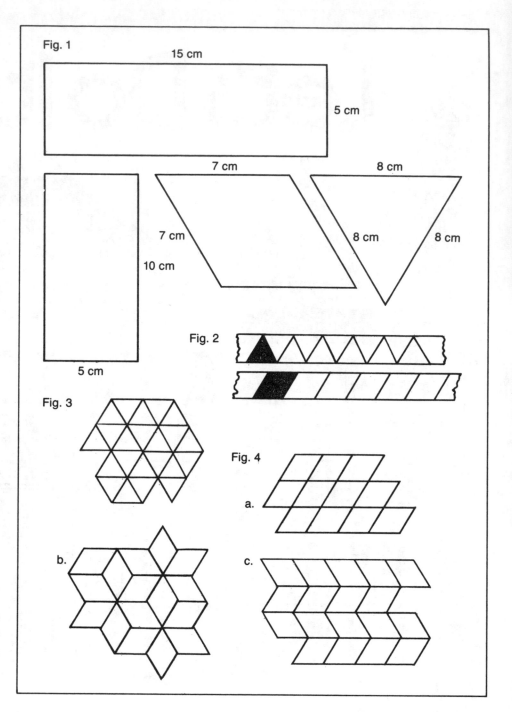

idea of a pattern is approached from the standpoint of a bricklayer, you can explain that once a wall or sidewalk pattern is begun, the pattern is simply continued in an orderly fashion.

Provide children with the cardboard tiles after discussing repeated brick patterns. Children can work in pairs or individually. Unless the class is relatively small and lots of table or floor space is available, this activity is probably not appropriate for a full class. Instead, after a general discussion, set up table or floor space where children can work in small groups.

With the cardboard tiles the purpose

is to establish the concept of a repeated pattern. The children should work experimentally with the tiles, trying to make various patterns and checking the regularity of each.

The equilateral triangle is virtually foolproof. There is only one way (fig. 3) that the triangle tiles can fit together as long as the corners meet. We included this triangle because it is an easy shape to work with. Later, when color is added, this tiling pattern becomes even more interesting.

The diamond (rhombus) is really two equilateral triangles put together. Several different patterns can be made

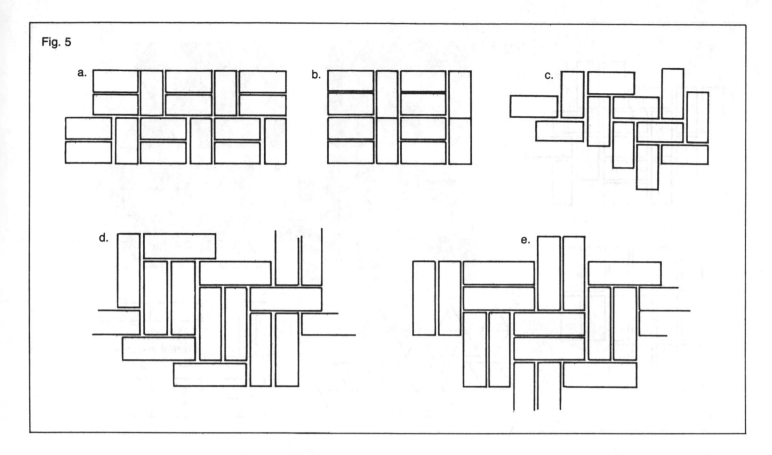

Fig. 5

with the diamond tile. Three such patterns are shown in figure 4. Note the straight rows of diamonds in figure 4a. This pattern is easier to see than the other two and is a useful one to show children who are having difficulty finding a pattern.

The two rectangular tiles have lengths and widths in the ratios of 2 to 1 and 3 to 1 respectively. Both provide virtually limitless variations and so permit greater creativity (fig. 5). As children are experimenting with the tiles, help them describe their patterns verbally to fix images of the patterns in their minds. In 5a, for example, the pattern could be described as "two and one and two and one . . . ," while pointing to each horizontal row. If the shift is varied, the pattern is changed to the one in 5b.

Many children will string out the tiles into long lines or create designs that are not tiling patterns because they have holes in them. Such designs may be fun, but they are not tiling patterns. Children who make such patterns need to be reminded of the notion of building a wall or tiling a floor (one with no boundaries). Walls and floors do not have holes. You may have to help some children by starting

a pattern for them and letting them continue it.

Many children will begin by building random patterns that do not continue in an orderly fashion (fig. 6a). Others will build very simple patterns (fig. 6b). Something good can be found in each. You can select a portion of a random design that begins a pattern and show children how to continue with their own "great idea." The child with the unimaginative design may be encouraged by noting how nice and neat everything is or by suggesting a slight variation, such as changing every other row so that the bricks lie in opposite directions.

This initial experimenting phase of making tiling patterns can become a spare-time activity for children over a period of one or two weeks. Keep all of the carboard tiles in boxes, one shape to a box, and allow everyone an opportunity to make a tessellation with each of the suggested shapes.

Making Permanent Patterns with Color

The cardboard tiles are nice for experimentation with geometric patterns but

patterns using colored paper are more appropriate for permanent designs. After children can create one or more geometric patterns with the cardboard tiles and they seem to comprehend the notion of repeating pattern, they should be given the opportunity to incorporate color into their tessellations.

To make permanent and colored patterns, the tiles are cut from colorful construction paper. They can be the same size as the corrugated cardboard tiles used for the initial patterns. These colored tiles are then arranged in patterns and glued or pasted on large sheets of unlined paper.

To make large numbers of tiles we have found it easiest to use a spirit master and duplicate the outlines of the tiles directly onto the construction paper. To save cutting time and paper, draw each tile repeatedly on a spirit master. Then feed the construction paper through the duplicating machine carefully, perhaps one sheet at a time.

Once the patterns are on construction paper, many children can cut out their own tiles. For younger children who do not have the necessary cutting skills, precut the tiles by using a paper cutter. Place five or six sheets of blank paper under one with the patterns du-

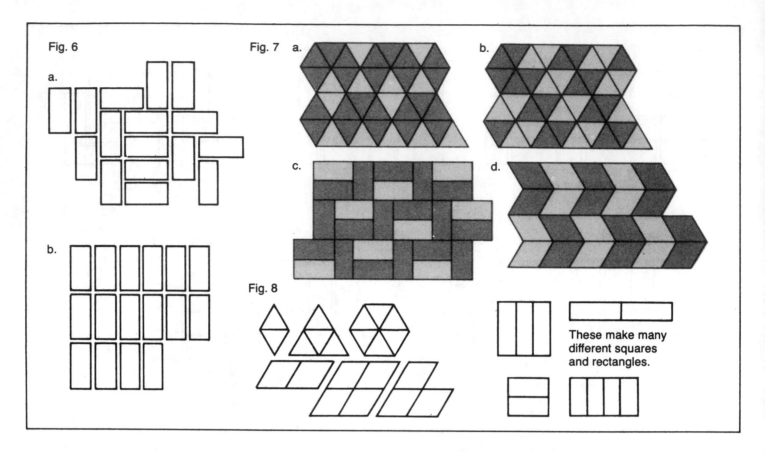

Fig. 6
a.
b.

Fig. 7
a.
b.
c.
d.

Fig. 8

These make many different squares and rectangles.

plicated on it. Then cut along the duplicated lines with the paper cutter, cutting all the sheets of paper at one time.

Give each child two colors of paper tiles of one shape. Each child should lay out her or his entire pattern on a large sheet of unlined paper before doing any pasting. In guiding the children, first be sure that their tiles form a repeating geometric pattern. Next, encourage the children to incorporate color into their patterns. One successful strategy is to have children make a geometric pattern using tiles of one color and then have them replace every other tile (or every third tile, or whatever pattern seems appropriate) with a tile of another color. The color patterns should be simple and repeat regularly in the same way that the geometric pattern repeats. A checkerboard is a simple example of a color pattern imposed upon a geometric pattern. With the four tile shapes suggested in this article, the color pattern possibilities are endless.

It is helpful to keep the number of tiles in one complete repetition of a color pattern rather small. Suggest ideas such as "make every other tile green", or "make the up-and-down

tiles red and the sideways tiles blue." Note that tiling patterns with the same geometric design can have many different color patterns imposed upon them. Thus even the triangle designs can be colored in many different patterns (figs. 7a and 7b). The problem-solving task of laying out a complete tiling pattern with repeating color design is a significant one. Many children will have difficulty, but all can be helped to develop a simple and pleasing design.

Be sure to have children get the colored tiles laid out completely before any pasting is done. Since the four tiles suggested for use here are symmetrical, a pattern will not change when each tile is turned over. Thus you can apply glue to the top side of about ten tiles in one section of the design, then turn each of these over and affix it permanently in place. This gluing method will minimize the problem of gluing a tile in an incorrect position.

Encourage the children to start near the center of their papers and work outward. Do not try to use the edge of the large sheet of paper as a guide. Work on getting the tiles straight with each other. If uneven outside edges result, these can be trimmed off.

Educational Payoff in Patterns

As noted, the problem solving involved in creating the patterns is significant. This feature alone justifies the time invested. However, the nice feature of these designs is that they can continue to generate learning experiences after they have been completed and displayed prominently as decorations in a classroom. The activities described here are all appropriate in the primary grades. Most can be done even in kindergarten. Those involving area, length, and symmetry are also profitable with upper-level classes.

Number and counting

Since the patterns are full of shapes made by small sets of tiles, they provide for lots of counting. Ask questions like these: "How many in this row (or square or group or triangle?) How many colored ones? Can you find a set of eight diamonds? Are there more here (point) or over there?"

Shapes within shapes

Two or more tiles frequently produce other shapes, depending on the patterns made. Figure 8 illustrates a few of the many possibilities. Some possible questions: "Can you find some tiles that make a square? a large triangle? a hexagon? How many different shapes can you find that are made with six tiles?" (See also fig. 3) Note that shapes do not need to be the familiar geometrical shapes. Let children describe or sketch any of the odd or unusual shapes they may be able to find in a design.

Measuring distances

The tiles we have suggested either have all edges the same length (diamond and triangle) or, as with the rectangles, the long sides are multiples of the short sides. Thus the edges of the tiles make excellent units of length for a given design picture. Then ask, "If you had to walk only on the cracks between the tiles, how far is it from this point to that point?" Try different paths and compare distances along each. It also is interesting to find all the corners in a design that are a given distance from a point near the center of the design.

The distance around a shape is a measurement properly called the perimeter. You could ask, "What shapes can you find that are exactly 10 units around?" Note that in the rectangle patterns, the longer edges must be counted as either 2 or 3 of the shorter units. If the longer edge is taken as one unit, the shorter edges are either one-half or one-third of a unit.

Area measurements

The tiles, arranged with no holes or gaps, provide an ideal model for teaching that area is a measure of covering. Trace or draw around a small set of tiles in a pattern. "How could we tell how big this shape is?" (Count the tiles) Compare several shapes within a design. "Which is bigger? How can we tell?" Try to compare long skinny shapes with fat ones. Look for shapes with specified sizes. "How many different shapes can we find that have an area of six rectangles?" Later use two or more tiles as the designated unit of measure. For example, three of the long rectangles from a square which makes a nice unit of area. "If this large square is one unit, what is the area of

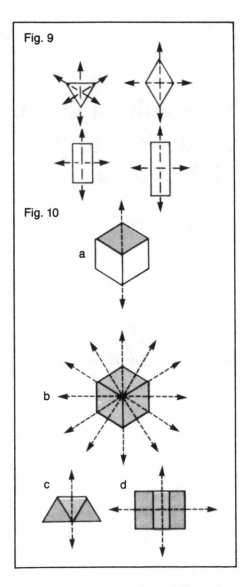

Fig. 9

Fig. 10

this whole design?" Use of larger units not only leads into possible fractional measures of area but allows the teacher to discuss how a unit can be used to measure a shape even when the unit does not "fit" into the shape.

Area and perimeter together

Questions on area or on perimeter can be combined to produce some interesting problems. "Find as many shapes as you can in this design that have a perimeter of 16 units. Sketch each one and record its area using one tile as a unit. What do you discover?" Or change this around. "Find as many different shapes as you can with an area of 8 units. What is the perimeter of each of these?" Next, combine these for a really hard search. "Find a shape in this design that has an area of 5 square units and a perimeter of 8." Questions such as these are not only

fun but they also dramatically contrast the often-confused concepts of area and perimeter without the use of formulas.

Symmetry

Each of the four tiles we have selected for use in these activities has at least two lines of symmetry (fig. 9). The tiling patterns invariably produce many more shapes with one or more lines of symmetry. Point to a design made of two or more tiles. "Where are the lines of symmetry in this design?" Next have searches for special cases of symmetry. "Who can find a design with four lines of symmetry? What is the largest symmetrical design in this tiling pattern? What design has the most lines of symmetry?" Figure 10 shows several small designs with lines of symmetry indicated. The colors involved should be included in consideration of symmetry. Therefore, the hexagon in figure 10a has only one line of symmetry but the same shape in figure 10b has six lines of symmetry.

Let's Do It

In this article we have outlined a procedure for placing tiling-pattern activities within the grasp of even the youngest school children. Using a collection of cardboard tiles made from the four basic shapes—equilateral triangle, rhombus, and rectangles of two different sizes—children can engage in rich experiences of spatial problem solving and shape awareness. Even use of the carboard tiles alone provides long-lasting, independent activity.

As color is added and discussions about resulting designs are conducted, the entire project begins to provide many more benefits. Geometric concepts come from the children's own work and provide real motivation. Art and mathematics come together to decorate the walls in a classroom.

A good bit of work goes into the making of materials for a tiling project such as the one described, but the ultimate payoffs are worth the effort; and the carboard tiles can be used over and over again. Maybe some parents will be willing to help make "tiles."

Why not get started today! ▰

Pictures, graphs, and transformations a distorted view of plane figures for middle grades

MARC SWADENER

An assistant professor of mathematics education at the University of Colorado in Boulder, Marc Swadener directs undergraduate and graduate programs in mathematics education. He has had experience as a secondary school mathematics teacher and has conducted inservice programs for both elementary and secondary school teachers. He has been a speaker at both state and national meetings of mathematics educators.

With the advent of the "new" mathematics movement about a decade ago, several new topics and concepts have been incorporated in school mathematics. The common example of this is the presentation of the idea of set at all levels of school mathematics. The introduction of new concepts has been on-going; a concept that has recently gained recognition is transformations on the plane.

Most of the literature on transformations, however, is geared for the secondary level, thus leaving the elementary and middle schools somewhat in the dark, but teachers at these lower levels can introduce concepts associated with transformation on the plane on an intuitive level that further enhances student understanding of mathematics in general.

One way to do this is by using pictorial graphs. These not only draw attention to transformations but provide further practice in working with coordinate systems, counting skills, and visualization of geometric shapes. They are also different from what students usually see in the classroom. The procedure that is proposed here

results from varying three things on the standard rectangular coordinate system:

(*a*) the units of measure on the axes;

(*b*) the angle between the axes; or

(*c*) the linearity of the axes themselves (that is, the axes are curves rather than straight lines).

The process is best explained through a series of graphs.

The basic picture is shown in figure 1 where it is graphed with standard rectangular coordinates. In the succeeding figures transformations are performed on this picture.

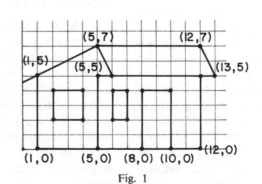

Fig. 1

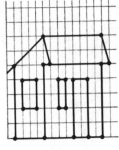

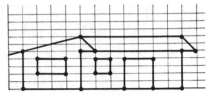

Fig. 2

Fig. 3

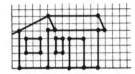

Fig. 4

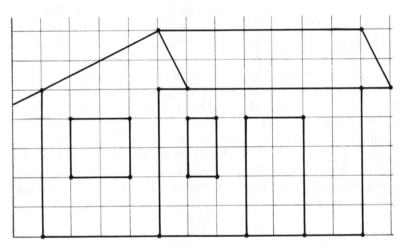

Fig. 5

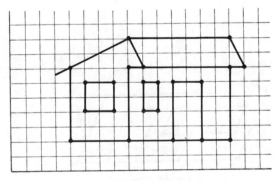

Fig. 6

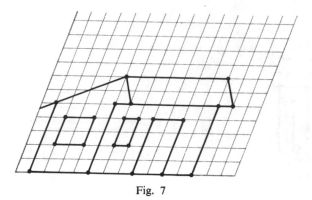

Fig. 7

In figure 2, the picture is represented with the horizontal axis changed; each unit on this axis is now one-half what it originally was. This corresponds to a shrinking (*contraction*) transformation on the horizontal axis with the vertical axis remaining constant. Figure 3 is a transformation of figure 1 by the same procedure except that the contraction this time is on the vertical axis, with the horizontal axis remaining constant. The transformations in figures 2 and 3 are sometimes called *one-way-stretches*.

Figure 4 exhibits a combination of the two previous transformations to contract the total figure. That is, a contraction by a factor of 1/2 has been performed on both the vertical and the horizontal axes. This corresponds to a *dilation* transformation with the dilation factor in this case being equal to 1/2. Figure 5 illustrates the results of dilating figure 1 using a dilation factor of 2 which doubles the length of each segment in the figure. This transformation is sometimes called an *enlargement* or *magnification*.

Figure 6 illustrates shifting the picture in figure 1 to the right three units and up two units. The shape and dimensions of the figure do not change in this case. The only thing that has changed is the position of the picture on the grid. This transformation is an example of a *translation*.

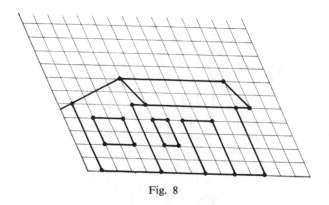

Fig. 8

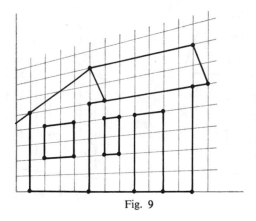

Fig. 9

Figure 7 changes the angle between the axes in the first quadrant to less than 90°. Figure 8 illustrates a corresponding change except that the angle between the axes (in the first quadrant) is greater than 90° but less than 180°. In both figures 7 and 8 the horizontal and vertical scales are uniform.

Figure 9 changes the original picture by using an expanding vertical scale as one moves away from the origin. The shape of the figure has changed. It should be pointed out that the horizontal lines in the grid, if extended to the left, would pass through a single point, commonly called a vanishing point in perspective drawing. The axes are perpendicular and the scale on the horizontal axis is uniform.

Figure 10 is very much like figure 9 except that the vertical scale gets smaller as you proceed away from the origin in a horizontal direction. This time the horizontal lines of the grid would pass through a single point if they were extended to the right. Figures 9 and 10 present transformations that are not ordinarily considered in schools. Interest could be generated in classrooms by having students name these transformations.

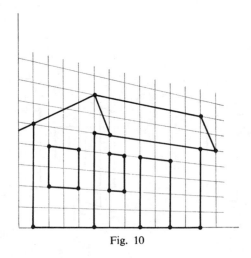

Fig. 10

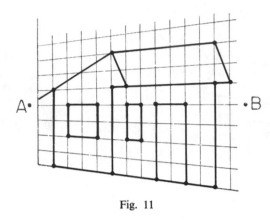

Fig. 11

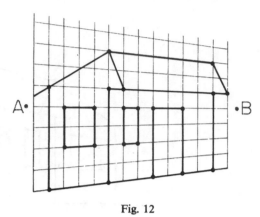

Fig. 12

Figures 11 and 12 are like figures 9 and 10 except that the line through *A* and *B*, instead of the horizontal axis, is perpendicular to the vertical axis. Again this type of transformation is not usually dealt with in schools. As in figures 9 and 10, the horizontal lines in the grid, if extended to the left in figure 11 and to the right in figure 12, would pass through a single point which corresponds to a vanishing point in perspective drawing.

Figures 13 and 14 combine properties of figures 11 and 12 to produce some interesting results. Figure 13 appears to be a perspective drawing (but isn't in actuality) and figure 14 is what could be called its opposite. These promote interesting discus-

sion among students since figure 14 just "doesn't look right."

In figures 15 and 16 the horizontal axes have been changed from straight lines to curves. In the first case the vertical lines are all parallel; in the second they radiate from a point on the vertical axis. Both figures 15 and 16 are quite different from what is usually seen.

In figure 17 both axes are curves (arcs of circles). The resulting figure is comparable to going to the "hall of mirrors" at a carnival. Once again this is a transformation not ordinarily dealt with in school.

In figures 2 through 17, the changes that made the distorted picture were uniform changes. There is, however, nothing sacred

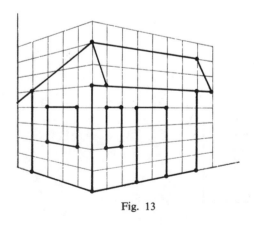

Fig. 13

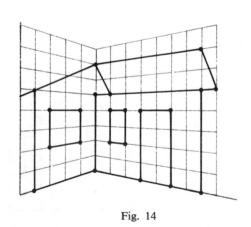

Fig. 14

GEOMETRY FOR GRADES K–6

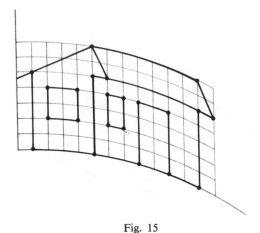

Fig. 15

about being uniform, as is illustrated in the last figure, figure 18. Here the coordinates are random and prove quite unusual. An example of such a transformation would be the "funny" mirrors at a carnival.

Transformations such as the ones dealt with here are easily presented in the classroom. A teacher could use them within a unit on graphing and include questions such as, "What would happen to the figure in the graph if the angle between the axes were less than 90°? Greater than 90°?" Questions like these would encourage students to mentally visualize the resulting changes and then "guess" the outcome. Or the material could be integrated in the year's program and individual aspects of these transformations could be presented where they would best fit into the total mathematics program.

Material like this should not be introduced at the elementary level for the purpose of teaching the specific transformations, except in very unusual situations or with a mathematics club of some kind. Why then would such things be done in a mathematics class? By the introduction of transformations in this way in the middle grades, students can see some unusual things happen and they can be introduced

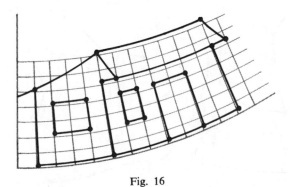

Fig. 16

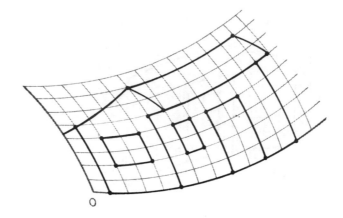

Fig. 17

to some of the terminology which they will later encounter. Most importantly, if students are allowed to identify for themselves the characteristics of the "changed" drawings they will be gaining experience in looking for out-of-the-ordinary things, seeing things from an entirely different standpoint. If we can educate students to see things from different perspectives we have done a notable service to mankind.

Presenting the actual material can be done in laboratory sessions, on worksheets, in small group discussion, or by other methods. It lends itself nicely to many approaches.

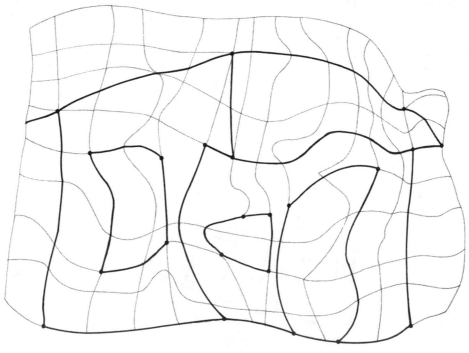

Fig. 18

Geometry in Unusual Ways

THESE articles are included because they describe somewhat unusual approaches to helping children develop an awareness of, and appreciation for, the place of geometry in their world.

The first two articles challenge students through activities involving cutting and folding. Edwards ("Discoveries in Geometry by Folding and Cutting") starts with a sheet of colored construction paper. By folding and cutting he gets two congruent right triangles. A series of foldings and cuttings of the triangular shapes follows, resulting in other shapes. In "Snipshots," Ranucci folds pieces of paper in various ways, makes some snips with scissors, and then asks students to predict what they will see when the paper is unfolded.

In "From Shadows to Mathematics," Bruni and Silverman have children make shadows with oak tag or cardboard shapes, and then make shadows with the shadows. These activities enable children to see shapes in different orientations and give them experience in systematically making and recording observations.

Many worthwhile geometric experiences can be provided inexpensively with readily available or teacher-made materials. Bruni and Silverman ("Using Geostrips and 'Angle-fixers'") show how to make the materials—geostrips and angle-fixers—and then describe ways they can be used in teaching. Lindquist and Dana ("Strip Tease") use strips of colored paper, a centimeter ruler, gummed tape, and scissors to make geometric shapes. In "Track Cards: A Different Activity for Primary Grades," Van de Walle describes an activity with easily made cards that encourages creativity and verbalization.

In "Frame Geometry: An Example in Posing and Solving Problems," Walter uses geometric ideas (rectangles) in a problem-solving context. The frames can be strips of wood or cardboard. Wahl describes an activity, "Marshmallows, Toothpicks, and Geodesic Domes," in which children build a three-dimensional figure. Although she focuses on geodesic domes, her idea could be applied to simpler three-dimensional figures as well. Modeling clay could be used in place of marshmallows.

Richardson ("The Möbius Strip: An Elementary Exercise Providing Hypotheses Formation and Perceptual Proof") challenges fourth-grade children with a Möbius strip (made with calculator tape and masking tape) and a Paul Bunyan tale. This is another example of problem-solving experiences with geometric materials.

Most schools have overhead projectors. In "Mathematics for the Overhead Projector," Arledge describes two games that a whole class can play with an overhead projector and some geometric shapes.

There are two special sets of geometric shapes that are easily made (commercially manufactured sets are also available) and make possible a wealth of geometric experiences. One is a set of tangrams, seven pieces made by partitioning a square in a prescribed way. Russell and Bologna ("Teaching Geometry with Tangrams") describe how the pieces can be made and then suggest ways they can be used for teaching. The second special set consists of twelve different arrangements of five squares. Pentominoes, as they are called, can be "discovered" by children and made from ordinary graph paper or bought as a game known as "Hexed." Pentominoes can be used for problem-solving and tiling activities, as described by Cowan in "Pentominoes for Fun and Learning."

Geoboards and Miras are relatively inexpensive classroom aids and can be bought from suppliers of instructional materials. Woodward introduces the Mira in "Geometry with a Mira." Teachers who are interested can build on his introduction or seek other references. He lists several.

The geoboard is a versatile instructional aid and besides being purchased can be made by anyone who is handy with hammer and nails. Every school should have at least one set. In "Geoboard Activities for Primary Grades," Kratzer and Allen describe a set of cards that can be made for use with a geoboard. Children are asked to reproduce on a geoboard the patterns shown on the cards.

Geometry is all around us. We can see it if we look. In "Let's Take a Geometry Walk," Nelson and Leutzinger suggest ways of helping children develop an awareness of geometry in their immediate environment. In "Some Geometry Experiences for Elementary School Children," Moulton calls attention to existing, practrical applications of geometric shapes. Children can see how geometric ideas are used in significant ways.

Geometry is a pleasure to teach. It is a subject that children can approach in their own individual ways. Children can make their own discoveries and often those who have difficulty with computation will succeed in geometric activities. Geometric materials can be used to develop many different mathematical concepts. In "Problem Solving with Five Easy Pieces," Lindquist shows the numerous ways a set of five geometric shapes can be used to solve problems in a variety of modes. Although teaching geometric concepts is not the purpose of the article (the geometric shapes are a means to a different end), handling these materials helps children become more familiar with the shapes and their distinctive chartacteristics.

Computers, too, can be a medium for children to discover geometric concepts. Craig ("Polygons, Stars, Circles, and Logo") describes three computer activities that give children another kind of hands-on experience with geometry.

Discoveries in geometry by folding and cutting

RONALD R. EDWARDS

An associate professor of mathematics at Westfield State College in Westfield, Massachusetts, Ronald Edwards teaches both undergraduate and graduate courses in mathematics and mathematics education. He also serves as coordinator of graduate programs in mathematics education at the college.

Most elementary students have had some experiences in paper folding—either with origami or such simple activities as folding a sheet of paper in quarters or constructing a paper airplane. Also, by cutting and folding, students may have constructed cubes, open boxes, snowflakes, or chains for Christmas trees. Students show interest in such activities and the following series of simple folding and constructing activities may prove both pleasurable and mathematically rewarding. Through these activities, relationships of geometric shapes and regions may be examined; definitions in geometry, reviewed; statements of geometric theorems, visualized; and students' creative talents, exercised.

MAKING AND FOLDING THE TRIANGLE

Begin with a sheet of colored construction paper (about 23 cm by 30 cm). Fold corner *A* so that edge *DA* lies along edge *DC* (fig. 1) and crease the paper sharply at this fold.

Opening the paper again, fold corner *C* so that edge *BC* lies along edge *BA*. Again, crease the fold and open the paper. Tearing or cutting along the creases *DE* and *BF* (fig. 2) produces two isosceles, right-triangle shapes and one parallelogram shape.

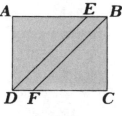

Fig. 2

Using one of the triangular shapes, make the series of folds indicated in figure 3, creasing the paper sharply for each fold. Note that each fold divides a triangular region in half. The completed folds separate the original triangular region into eight congruent triangular regions.

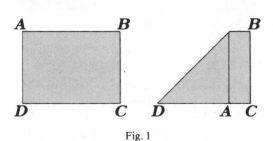

Fig. 1

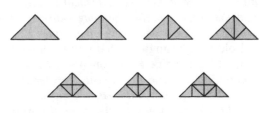

Fig. 3

SOME FOLDING ACTIVITIES

Fold your triangular region along the creases so that a region of each of the following shapes is formed:

1. a square
2. another square (either larger or smaller than the first)
3. a rectangle (that is not a square)
4. a triangle whose area is 1/2 that of the original region; 1/4 that of the original region; 1/8 that of the original region
5. a parallelogram (that is not a rectangle)
6. a trapezoid
7. another, different trapezoid

SOME GEOMETRIC THEOREMS

Place your triangle so that it is positioned as shown in figure 4. Note that *the longest side of the triangle is opposite the angle of largest measure.* In a right triangle, the side opposite the right angle (the largest angle) is called the hypotenuse.

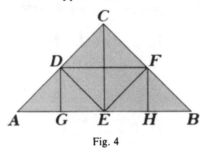

Fig. 4

Fold your triangle so point B lies on point E, point A lies on E, and point C lies on E as in figure 5. Note that *the sum of the measures of the interior angles of the triangle is 180 degrees.*

Fold your triangle so point B lies on point A. Segment CE is the altitude to side AB and $\angle A \cong \angle B$. Note that *the altitude from the vertex angle of an isosceles triangle divides the base into two congruent segments* and *the sides opposite the congruent angles of an isosceles triangle are congruent.*

Fold your triangle so that point A lies on point C and point B lies on point C. Segments CE, AE, and BE are the same length. Note that *the midpoint of the hypotenuse of a right triangle is equidistant from the three vertices of the triangle.*

These are examples of theorems that can be visualized using your folded triangular regions.

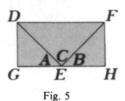

Fig. 5

SOME CONTRUCTION ACTIVITIES

Cut your triangular region along the folds, giving you eight congruent, triangular regions. In the following activities you will use these small triangular regions to build more complex regions. However, the triangles must be joined in one of the three ways indicated in figure 6.

Fig. 6

1. Exchange four small triangular regions with someone else, giving you four of one color and four of another. Form patterns of the type illustrated in figure 7.
2. Using eight triangles, form a rectangle that is not a square.
3. Form another, different rectangle.
4. Using six triangles, form a hexagon.
5. Experiment with forming other shapes.

Fig. 7

FURTHER INVESTIGATIONS

Using two of the small triangular regions, there are only three different geometric shapes that can be formed (fig. 6)—a square, a triangle, and a parallelogram.

Others may look different, but they are merely rotations or reflections of these three.

1. How many geometric shapes can be formed by joining three triangular regions?
2. Four triangular regions?
3. Five triangular regions?
4. Keep a record of the possible shapes you find in the investigations 1, 2, and 3.

SOME ACTIVITIES WITH THE SCRAPS

Not included in the foregoing activities is the parallelogram shape region, labeled *DFBE* in figure 2. The following are a few activities you may try with these scraps.

1. Without using a pencil, ruler, or compass, fold the parallelogram shape as shown in figure 8, where each triangle that is formed is isosceles and all triangles are congruent to $\triangle ABC$. In $\triangle ABC$, $AB \simeq AC$. Cut along fold *HI* and discard the region *HIJK*. Join the ends of *ABHI* so that point *I* lies on point *C*, point *H* lies on point *B*, and point *G* lies on point *A*. Glue or staple these two regions together forming a three-dimensional shape.

2. Make several more of these three-dimensional shapes and try joining them to make three-dimensional structure.

3. From the parallelogram region you can construct a hexaflexagon (Berger 1951, Olson 1975).

4. Using the parallelogram region explore the Möbius strip or paper knots (Johnson 1957, Olson 1975, Richardson 1976).

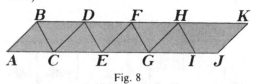
Fig. 8

References

Berger, Emil J. "Devices for a Mathematics Laboratory." *Mathematics Teacher* 44 (April 1951):247–48.

Johnson, Donovan. *Paper Folding for the Mathematics Class.* Reston, Va.: National Council of Teachers of Mathematics, 1951.

Olson, Alton, T. *Mathematics through Paper Folding.* Reston, Va.: National Council of Teachers of Mathematics, 1975.

Richardson, Lloyd I. "The Möbius Strip: An Elementary Exercise Providing Hypotheses Formation and Perceptual Proof." *Arithmetic Teacher* 23 (February 1976):127–29.

Fig. 1

a

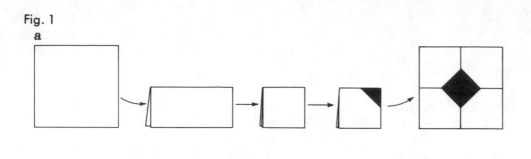

b

c

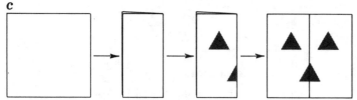

d

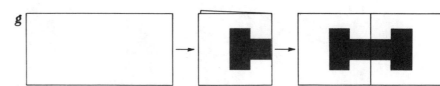

e

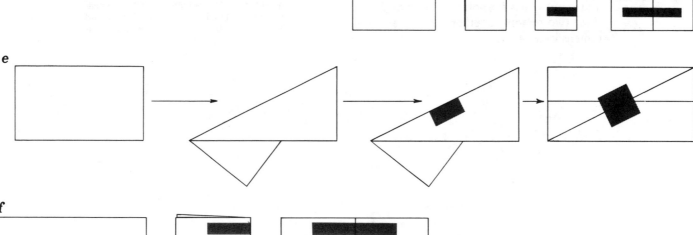

f

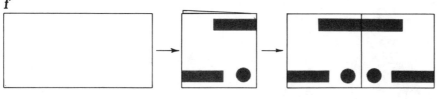

g

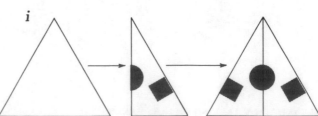

h

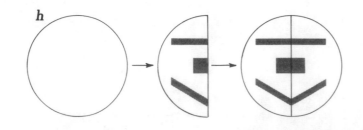

i

Snipshots

By **Ernest R. Ranucci**

The *snip* in the title is no error of the printer. It refers to the snip of a pair of scissors. The materials needed for the lesson I am about to describe are simple: a pair of scissors, some sheets of paper, and a blackboard. (*I know.* You call it a chalkboard and it's sometimes green—so what else is new!)

You hold up a rectangular sheet of paper, fold it in some manner, then cut it once or twice. The children are then asked to imagine what the paper will look like after it's opened. They then sketch a picture of the perforated sheet—some of them do it at the blackboard. Simple? Yes, but a lesson of this type offers rich possibilities for creativity. What's more, it is applicable to students of varying ages and abilities. It can be used anywhere from kindergarten to grade eight and, with suitable complexity, at high-school levels.

For the kindergarten and primary grades, I usually begin as follows, describing the folding as I do it:

Here we have a sheet of paper. I'm going to fold it from top to bottom. What do you suppose it would look like if I open it?

Draw a picture of the paper and its crease.

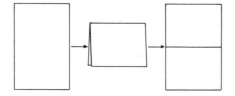

Let's do the same thing, but this time I'll continue by folding from left to right. Draw me a picture of what it will look like when it's opened.

This time I'll fold the paper from left to right. Then I'll fold it once more from left to right. Then I'll open it. What do you suppose it will look like?

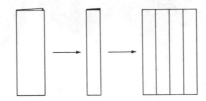

Suppose we repeat what we just did, but after the second fold we'll fold the entire thing from top to bottom. Draw me a picture of the opened sheet.

That's just for a start. Try different folds, creasing the paper into thirds, along one diagonal, along two diagonals, into thirds followed by a crease along one diagonal, and into thirds and sixths. When you feel that the students are ready for more complex maneuvers, try cutting opening in the folded sheet. I'll describe the first of the perforations in some detail. After that, let your intuition be your guide; just follow the drawings. (Fig. 1) Invent your pattern as you go along.

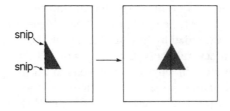

I'm going to fold a sheet from left to right. Then I'm going to cut two short snips along the folded edge (a piece of paper falls to the floor). Can you draw me a picture of what the sheet will look like when we open it? Don't forget to draw in all creases. □

Prior to his untimely death in August 1976, Professor Ranucci was on the staff of the State University of New York, Albany. The manuscript for this article was submitted to the Arithmetic Teacher by Professor Ranucci himself just before his death.

GEOMETRY IN UNUSUAL WAYS

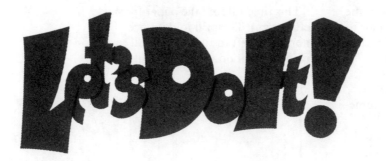

By **James V. Bruni** and **Helene J. Silverman**,
Herbert H. Lehmán College, City University of New York

From shadows to

Shadows are fascinating! They come in a wide variety of shapes and sizes. Sometimes you see them, sometimes you don't. When they do appear, you can't pick them up, nor can you weigh them. Shadows change shape and direction at different times of the day.

A child's natural curiosity about this phenomenon can lead to many exciting and unusual mathematical and scientific investigations. In addition to the realization that shadows are formed when light rays are blocked, there is so much more that children can find out about shadows.

Using different light sources, sunlight and then artificial light, some basic ideas about shadows are explored in the activities that follow. The central question, throughout the activities, is *How is the shape of an object different from or the same as the shape of its shadow?* Children are encouraged to apply concepts of number, measurement, and geometry as they explore and discover new ideas about these special kinds of transformations of shapes.

SHADOWS FROM SUNLIGHT

Children can easily begin experimenting with shadows by examining their own shadows outdoors on a sunny day. (See "Doing Things with Your Shadow.") They can try these activities at different times in the day (and at home) and begin to realize that their shadows vary in length at different times of the day. Older children can measure their shadows, measure their heights, make a chart with the results of their measurements, and begin to explore the relationship between the heights of people or objects and the lengths of their shadows. (See "Measuring Our Shadows.")

This activity can be extended by having children make "shadow clocks." Shadow clocks can be made in several ways. One easy way is to set a thin stick (a wooden dowel 20 to 30 centimeters in length works well) in an upright position. It can be stuck into the ground or into a piece of Plasticine on the pavement. Then have the children mark the position and length of the stick's shadow on each hour. Several shadow clocks (using sticks of different lengths) might be made. How will the clocks be different? How will they be the same? (See "Observing a Shadow Clock.")

If you have an area of the room that has sunlight for several hours of the day, you

mathematics

Photo by Clif Freedman

might want to make shadow clocks indoors. You can use a pencil positioned upright (with a small piece of Plasticine to hold it in place) on a large sheet of paper.

Using oaktag shapes

Make oaktag or cardboard shapes—different kinds and sizes of triangles, squares, rectangles, parallelograms, trapezoids, and so on. Label each shape for easy identification.

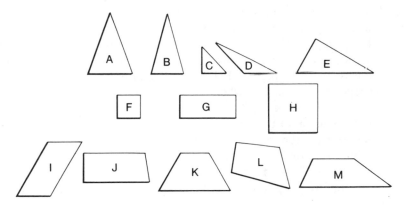

Punch two or three holes in some of the shapes. To make a handle, use tape to attach a piece of wire (an extended paper clip will do) to each shape.

These oaktag shapes can then be used for a variety of shadow activities. (See "Making Shadows of Shapes with Holes" and "Making Shadows with a Triangular Shape.")

The children can draw the shadows that they form with the oaktag shapes on a large piece of paper. It is best for the children to work in pairs, with one child

holding the shape and forming shadows while the other child outlines the shadows formed. The outlining is easier if the corners (or vertices) of the shadow shape are marked and then line segments drawn with a straightedge.

Pairs of children can make task cards with the shadow shapes that are formed by a particular shape. Then other children can try to form the shadows indicated on these task cards. (See "Making Shadows like Marcia's and Lillian's.")

The children can do this activity indoors by using the light from a slide projector as an approximation of the sun's rays. They can hang a sheet of paper (on which they project shadows) on a wall about four feet from the projector. The results will be very much like the shadows formed by sunlight.

They will be able to make many different kinds of triangular shadows from the triangular oaktag shapes. The shadows may have sides and angles with measures different from those of the original figures. But, certain things don't change: the number of holes doesn't change and three-sided figures always have three-sided shadows (except in the special case when you tilt the triangle in such a way that the shadow looks like a line segment).

An interesting extension of this activity is for the children to draw the shadows of a shape on oaktag or cardboard, cut out the outline of the shadow, and then make shadows from that "shadow" shape. Ask the children to try to make a shadow of the "shadow" shape that is the same size and shape as the original shape.

From shape to shadow shape and from shadow shape to shape is an example of an inverse operation.

Shape A $\xrightarrow{\text{has shadow}}$ Shape X then Shape X $\xrightarrow{\text{has shadow}}$ Shape A

It is similar to the inverse operations of adding a number and then subtracting that same number.

$$5 + 3 = 8 \text{ then } 8 - 3 = 5.$$

What kinds of shadows are formed by the four-sided oaktag shapes? Four-sided shadows are formed in each case, although the lengths of sides and sizes of angles may change. (See "Making Shadows with Four-Sided Shapes.") Many different

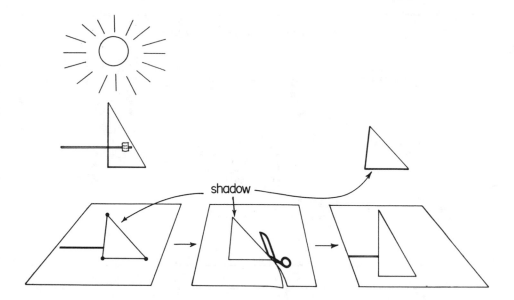

kinds of quadrilateral shadows can be formed when the oaktag shape is four-sided, but not *all* kinds of quadrilaterals can be formed with any particular four-sided shape. In sunlight the square oaktag region has shadows shaped like squares, rectangles, or parallelograms, but *not* trapezoids (with only two sides parallel) or quadrilaterals with no pairs of parallel sides.

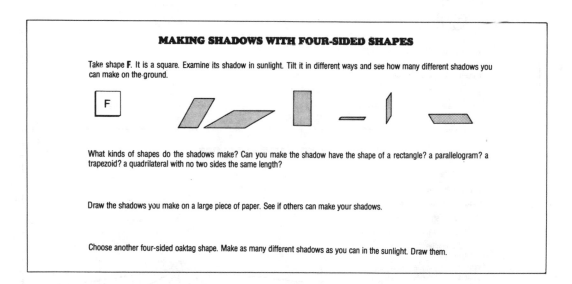

MAKING SHADOWS WITH FOUR-SIDED SHAPES

Take shape **F**. It is a square. Examine its shadow in sunlight. Tilt it in different ways and see how many different shadows you can make on the ground.

What kinds of shapes do the shadows make? Can you make the shadow have the shape of a rectangle? a parallelogram? a trapezoid? a quadrilateral with no two sides the same length?

Draw the shadows you make on a large piece of paper. See if others can make your shadows.

Choose another four-sided oaktag shape. Make as many different shadows as you can in the sunlight. Draw them.

Similarly, the shadow of the rectangular piece of oaktag can be shaped only like a square, a rectangle, or a parallelogram. The shadows of a parallelogram shape also will have to be shaped like a square, rectangle, or parallelogram. What is happening as the sunlight forms these shadows? The shadows of these shapes do *not* necessarily have sides and angles that àre the same size, respectively, as the original shapes.

DOING THINGS WITH YOUR SHADOW

Can you find your shadow?
Can you make your shadow move?

Can you make your shadow
stand on one leg?

How tall (or short) can you
make your shadow? How fat (or
thin) can you make your shadow?

Can you walk around without
having your shadow "bump
into" anyone else's shadow?

When you walk around, is your
shadow behind you? Can you make
it walk in front of you?
Can you "save your shadow"? Have
your partner trace your shadow on
a piece of paper (newsprint).
Then cut out your shadow.

Who has the longest shadow? Why?

MEASURING OUR SHADOWS

Date: **Oct. 10** Time: **2:00 p.m.**

Name	Your height	Length of your shadow
Michael	115 cm	170 cm
Cynthia	120 cm	179 cm
Richard	125 cm	188 cm

Who has the longest shadow? Why?

If someone here were 180 cm tall, about how long do you think that
person's shadow would be? Why?

OBSERVING A SHADOW CLOCK

This is a shadow clock.

Does the shadow change? How?

When is the shadow longest?

When is the shadow shortest?

Mark the lengths of the shadow
at different times.

When is the length of the shadow the
same size as the height of the stick?

When is it twice as long?

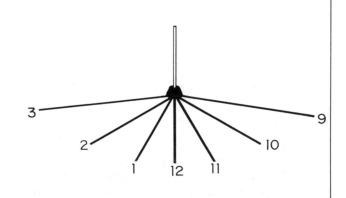

MAKING SHADOWS OF SHAPES WITH HOLES

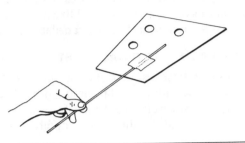

Choose an oaktag shape that has holes in it.
How many holes does it have?

Make shadows with your shape. Tilt the shape
in different ways to make new shadows.
How many different shadows can you make?

How many "holes" are in the shadow? Can you
make a shadow with fewer holes? more holes?

MAKING SHADOWS WITH A TRIANGULAR SHAPE

Take Shape **A**. What kind of shape does it have? Look at its shadow in sunlight. Tilt the shape in different ways. How many different shadows can you make on the ground?

Make the largest shadow you can. Make the smallest shadow you can.

Make the longest shadow you can. Make the shortest you can.

Which of the following shadow shapes can you make from Shape **A**?

1. square
2. right triangle
3. obtuse triangle
4. isosceles triangle
5. scalene triangle
6. rectangle

MAKING SHADOWS LIKE MARCIA'S AND LILLIAN'S

Names: *Marcia*
         Lillian

Shape used:

Shadows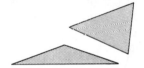

Can you make the shadows Marcia and Lillian made?

What does remain the same as the shape forms its shadow? *In sunlight,* pairs of sides that are parallel in the original shape always appear parallel in its shadow. A square, then, *can* have a shadow shaped like a square, rectangle, or parallelogram. The square *cannot* have a shadow shaped like a trapezoid, with only one pair of opposite sides parallel, or like some other quadrilateral with no pairs of parallel sides.

SHADOWS FROM ARTIFICIAL LIGHT

Experiences like those just described can be repeated indoors using an artificial light. A clear, incandescent bulb, a flashlight (with reflector covered with tape), or a high-intensity lamp work well. What kinds of shadows can be formed indoors with this kind of lighting? Will the shadows be different from the ones formed outdoors with sunlight?

As with shadows made by sunlight, the sizes of the sides and angles of the shadow shapes may be different from the corresponding sides and angles of the oaktag shapes, but something else changes, too. Now, using a square oaktag shape, you *can* make a shadow shaped like a trapezoid or a quadrilateral with no pair of parallel sides.

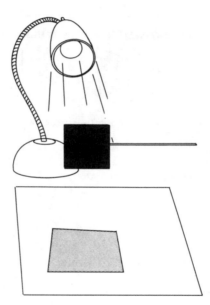

In the sunlight, when an oaktag shape had parallel sides, its shadow always had parallel sides, but in artificial light this is not necessarily so. An oaktag shape can have parallel sides and its shadows may have no parallel sides. This is because the light rays from the sun come from such a great distance that they seem to be parallel. Light rays from the bulb, however, are clearly not parallel—they fan out from the bulb.

Making similar shapes

When artificial light is used and the original shape is kept parallel to the surface on which the shadow is formed, shadows that are similar to the original oaktag shape (same shape but not necessarily the same size) can be made. One way to make shadows that are similar to the original figure is to attach a flashlight to the leg of a table so that the flashlight is pointed downward. Then hold an oaktag shape parallel to the floor and project its shadow onto a piece of paper on the floor.

Children can explore the relationship between the lengths of the sides of the oaktag shape and the lengths of the corresponding sides of its shadow. Since the shape and its shadow are similar (same shape but possibly different size), the measures of the lengths of the corresponding sides form the same ratio.

	Oaktag shape L	Shadow of L	Ratio of lengths of corresponding sides
side #1	5 cm	10 cm	$\frac{5}{10}$
side #2	6 cm	12 cm	$\frac{6}{12}$
side #3	7 cm	14 cm	$\frac{7}{14}$
side #4	8 cm	16 cm	$\frac{8}{16}$

By attaching a ruler or tape measure along the leg of the table, more sophisticated relationships can be investigated. For example, children can investigate the relationship between the distance of the shape from the bulb of the flashlight and the size of the shadow.

The activities described in this article only begin to suggest the many kinds of experiences children can have with shadows. Besides the measurement and problem-solving opportunities they provide, these experiences can help develop a better understanding of shape and, along with it, an intuitive background for more advanced mathematical topics.

More activities with shadows

Dienes, Z. P., and E. W. Golding. *Geometry of Distortion.* New York: Herder and Herder, 1967. Available from McGraw Hill, New York.

Elementary Science Study. *Light and Shadows.* St. Louis: Webster Division, McGraw-Hill Book Co., 1968.

Trivett, Daphne Harwood. *Shadow Geometry.* New York: Thomas Y. Crowell Co., 1974.

By **James V. Bruni** *and* **Helene Silverman,**
Herbert H. Lehman College, City University of New York

Using geostrips and "angle-fixers"

Geometry is an important, exciting part of elementary school mathematics. Children can be involved in relating mathematics to shapes in their environment. The activities described here suggest one way to help children develop some basic ideas about simple closed shapes, angles, and triangles. By constructing models of geometric shapes and physically transforming those models, children can examine the changes that occur with a transformation. This can help organize and synthesize thinking about basic geometric concepts. It can also serve as a foundation and a springboard for activities involving the use of a ruler, compass, or protractor.

The materials used in these activities are homemade geostrips, "angle-fixers," and brass fasteners, as shown in figure 1. (The angle-fixers are patterned after

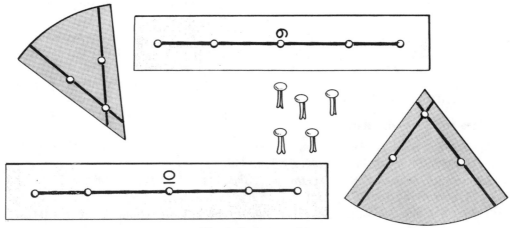

Fig. 1. Basic materials

Photograph by Clif Freedman

to develop ideas about shapes and angles

a material developed by Emma Castelnuovo, a teacher in Rome, Italy. The authors are indebted to Professor Castelnuovo for her pioneering efforts in the teaching of geometry through the transformation of models.) The geostrips and angle-fixers are made by using rubber cement to mount patterns like the ones in figures 2 and 3 onto oaktag and cutting along the black lines. The holes are made,

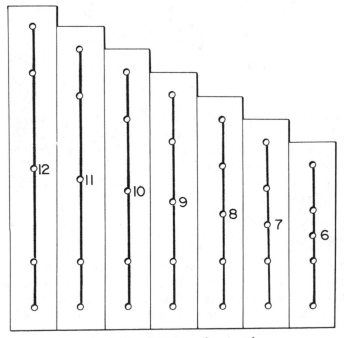

Fig. 2. Sample patterns for geostrips

as indicated, with a hole punch (preferably one that makes very small holes). It is convenient for the strips to be 2 centimeters wide and to have different lengths. The numbers indicated on the strips represent the distance in centimeters between the end-holes of the strips. In making models with the geostrips, this distance is the length. You will need strips ranging in length from 6 to 24 centimeters. Each strip has five holes: two holes that are 2 centimeters apart at each end of the strip and one hole at its midpoint.

To make the patterns for the angle-fixers, begin with a circle with a radius of 5 centimeters. Make "sectors" that vary in angle size. (Fig. 3.) The patterns for angle-fixers should include 30, 45, 60, 75, 90, 105, 135, 150, and 180 degree

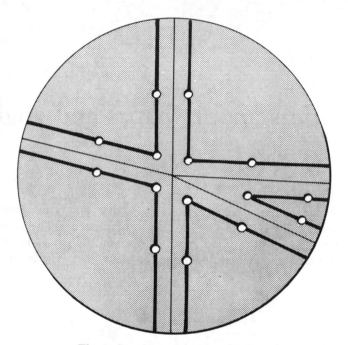

Fig. 3. Sample pattern for angle-fixer

angles, but initially they should not be labeled as such. You may wish to have the children label the angle-fixers in degrees later when you introduce the degree as a unit of angular measurement. On each sector draw a red line segment along each straight side ½ centimeter from the sides, as shown, and punch a hole where the segments intersect. Each of the holes along the "sides" of the angle-fixers should be 2 centimeters from this hole.

Simple closed shapes

Give children an opportunity to make different kinds of shapes using just the strips and brass fasteners. Encourage the children to talk about the shapes that are made: How are the shapes alike? How are they different? Once children have had ample opportunity to explore the materials, you might pose a challenge like the following.

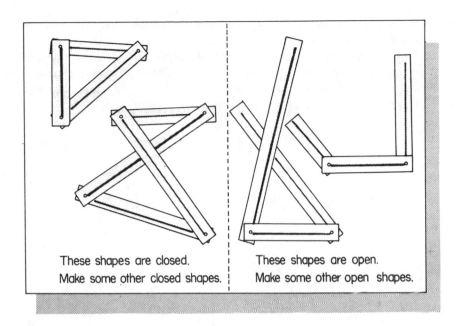

These shapes are closed.
Make some other closed shapes.

These shapes are open.
Make some other open shapes.

See if the children can discover the difference between an open shape and a closed shape from the examples. Let the children attempt to explain that difference in their own way. We are trying to make them realize that a closed curve has no "beginning" or "end." If a child moves his finger along the closed shape, he can keep going "around and around" along the shape. Once children can successfully make an open shape, ask them to make an open shape into a closed shape. When children can transform an open shape to a closed shape, and vice versa, you have a better assurance that they understand the concept. Notice that a shape can "cross itself" or not. That has nothing to do with being open or closed, which brings us to a second challenge. Again, when children can make a simple

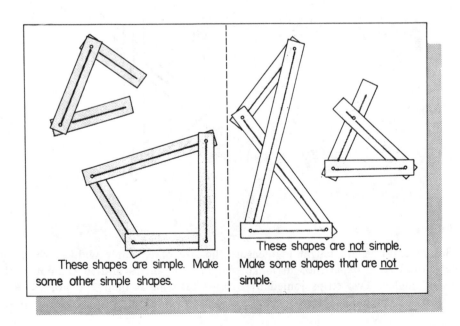

These shapes are simple. Make some other simple shapes.

These shapes are <u>not</u> simple. Make some shapes that are <u>not</u> simple.

shape, they should be asked to try to change the simple shape into one that is *not* simple, and vice versa.

Once children are familiar with the ideas of *simple* and *not simple* shapes, and *closed* and *open* shapes, they can try to make shapes that are both simple *and* closed at the same time. If each strip represents a line segment, these simple, closed shapes that are made up of line segments represent *polygons*.

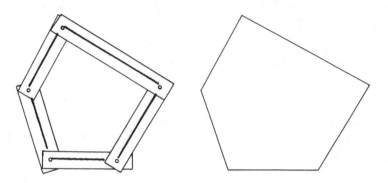

Shapes that are simple and open or not simple and closed, do *not* represent polygons. Children should be encouraged to transform these shapes that are not polygons into models of polygons.

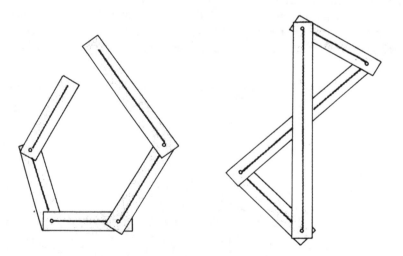

Angles

Before children are taught to measure the size of an angle with a protractor, they should have experience in making models of angles and in estimating the sizes of angles. Two strips joined by a brass fastener at their end holes can be thought of as a model of an angle. You can tell children to hold one strip still and rotate the other one about the brass fastener. Then ask, What is changing? Children should see that as one strip is turned the "amount of opening" between the strips changes. Ask the children, What is the largest opening you can make? What is the smallest opening you can make?

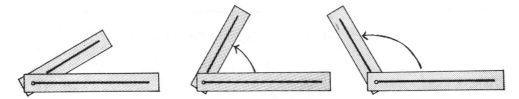

At this point the "angle-fixer" can be introduced. By using an angle-fixer you can keep the opening between two strips from changing; in other words, you can "fix" an angle. The children can use different angle-fixers and compare the resulting models.

Brass fasteners

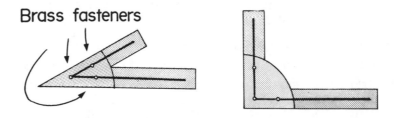

How can children show that one opening is larger than another? They can place one model on top of the other.

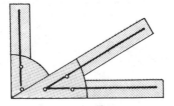

Children can also realize that the size of an opening does *not* depend on the length of the strips used. (How often do children confuse size of angle with length of "sides"?) As the children compare the sizes of openings, you might ask them to arrange the models in order from smallest to largest opening.

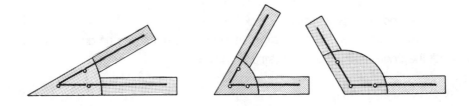

Once children realize how the angle-fixers influence the sizes of these models of angles, you can introduce activities such as the following to develop an understanding of specific kinds of angles.

Find an angle–fixer that makes an angle like the corner of a book.

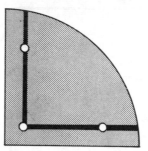

This kind of angle is called a right angle.
Make a model of a right angle using two geostrips and an angle–fixer.
Find all the angle–fixers that form right angles.

Find an angle–fixer that makes an angle like the edge of a book.

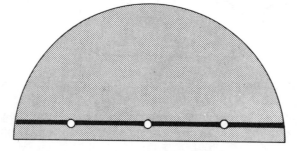

This kind of angle is called a straight angle.
Make a model of a straight angle using two geostrips and an angle-fixer.
Find all the angle-fixers that form straight angles.

When the right angle and the straight angle are understood, models of acute and obtuse angles can be made and identified.

Find all the angle-fixers that can form angles that are <u>smaller</u> than a right angle.

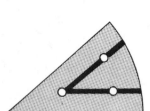 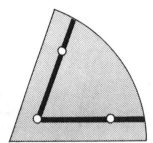

These angles are called <u>acute angles</u>.
Make some models of acute angles using geostrips and angle-fixers.

Find all the angle-fixers that can form angles that are <u>smaller</u> than a straight angle, but <u>larger</u> than a right angle.

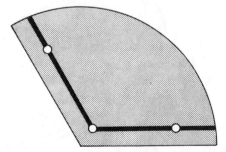

These are called <u>obtuse angles</u>.
Make some models of obtuse angles using geostrips and angle-fixers.

When you introduce the degree as a unit of angular measurement, the children can discover ways to find the measures of all the angle-fixers: Starting with an angle-fixer that can form the model of a right or 90° angle, children can find the number of degrees in the other angle-fixers. By placing smaller angle-fixers on top of the 90° angle-fixer, children can then figure out the number of degrees in the small angle-fixers.

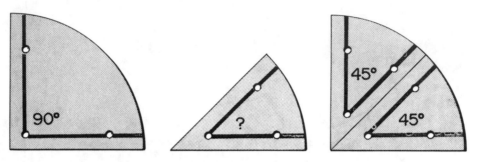

In a similar way the children can find the measure in degrees of all the angle-fixers without using a protractor. These kinds of experiences can give them a better understanding of angular measurement and increase their ability to estimate the size of an angle.

Triangles

In the previous activities the children found that a model of a simple closed shape made with the strips is a model of a polygon. They can discover that they need *at least* three strips to make a polygon. A three-sided polygon is called a triangle. Can children make a triangle with *any* three strips? An activity like the following can be interesting.

Close your eyes. Choose any three strips. Now open your eyes and make a triangle with these strips. Try again. Make other triangles.
Can you always make a triangle with any three strips?

The children should be encouraged to keep a record of when it is and when it is not possible to make a triangle with three strips. See what the children can discover about triangles from the chart.

Can You Make a Triangle?			
Side 1	Side 2	Side 3	Yes or No
6	7	8	Yes
6	7	15	No
8	9	14	Yes
8	9	20	No
7	18	6	No
8	15	10	Yes
9	18	7	No

As the children make many different triangles you can ask them how they might sort the triangles into three piles, placing all those triangles that seem to belong together into the same pile. Children may sort or classify the triangles in different ways; discuss with the children how they chose to sort the triangles. This discussion can lead to the introduction of the terms *equilateral* (three sides congruent), *isosceles* (two sides congruent), and *scalene* (no sides congruent).

The children can use the angle-fixers to discover facts about the angles of these three kinds of triangles: All three angles of an equilateral triangle are congruent. Two angles of an isosceles triangle are congruent. No two angles of a scalene triangle are congruent.

Children may also discover another way of classifying the triangles according to their angles. You might ask the children if they can make a triangle with two obtuse angles, or a triangle with two right angles. And why does every triangle they make have to have *at least* two acute angles?

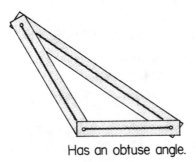

Has an obtuse angle.

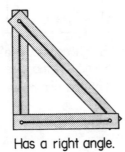

Has a right angle.

Has all acute angles.

If the children can classify triangles in terms of sides and in terms of angles you can try an activity like the following.

Make the following pairs of triangles with geostrips.

Sides: 6, 7, 8
         12, 14, 16

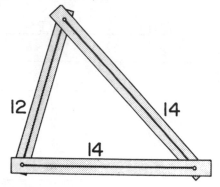

These triangles have the same <u>shape</u> although they are different sizes. They are <u>similar</u> triangles

What do you notice about the angles of these triangles? How are the lengths of the sides related?

Children can investigate the idea of similar triangles by making special pairs of triangles. Activities like the following can be used.

Each space below suggests a possible triangle, described according to its sides and according to its angles. For example, the triangle indicated is acute and scalene.

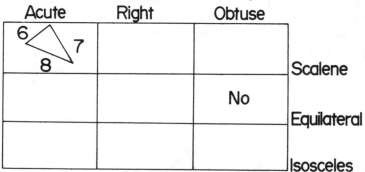

	Acute	Right	Obtuse	
Scalene				
Equilateral			No	
Isosceles				

Can you make one that is right and scalene? Can you make one that is acute and equilateral? Which kinds of triangles are possible to make? Draw a picture of each one that you can make. Indicate which ones cannot be made.

Make the following sets of triangles with geostrips.

A. 6, 8, 10
    12, 16, 20

B. 6, 6, 8
    10, 10, 11

C. 6, 6, 8
    18, 18, 24

D. 6, 7, 8
    9, 10, 11

Which ones are similar?

Can you make up some sets of similar triangles?

Investigating other shapes

You can encourage the investigation of many ideas about triangles or other polygons. For example, when the children make models of polygons with more than three sides, they find the models can be "pushed out of shape." (Fig. 4.) They cannot "push" a model of a triangle "out of shape," however; a triangle is a *rigid* figure. This makes the triangular shape very valuable for construction (as the shape of supporting structures).

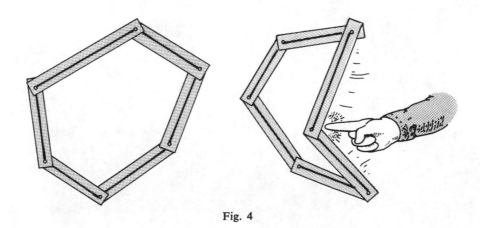

Fig. 4

The children can examine what happens when a polygon is "pushed out of shape." Notice that in this activity the perimeter remained the same as the area changed. Other questions can be raised: When is the area the greatest? How are the sizes of the angles changing?

Make a square with geostrips.

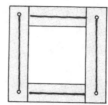

This is not a rigid figure. If you push in at one vertex, you transform the square into a rhombus like this:

Push →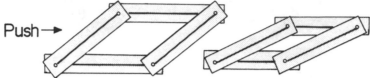

What has changed with this transformation? What remains the same? Are the opposite sides still parallel? Has the length of the sides changed? Does the perimeter change? The area? How about the size of the angles?

By using geostrips and angle-fixers to actually make models of geometric shapes and by transforming those models, children can gain a great deal of insight and understanding about basic geometric ideas.

Photograph by Clif Freedman

Let's Do It

Strip Tease

By **Mary Montgomery Lindquist,** *National College of Education, Evanston, Illinois, and* **Marcia E. Dana,** *Madison, Wisconsin*

What can be done in a mathematics class with strips of paper? Let this article tease your mind with ideas of how to use paper strips to make geometric figures, to practice measuring, to investigate patterns, to practice addition facts and reading numbers, and to explore other mathematical ideas.

The paper strips are easy to prepare, inexpensive, and versatile, but their real beauty is the ease with which children can make geometric figures. A child takes a strip, folds it, and tapes it as shown in figure 1. In the activities suggested in this article the children are often asked to measure a length before folding or to fold the strip into fractional parts such as thirds or fourths. Other than the acquisition of these skills, most of the activities are appropriate for any age level, depending on how you structure the directions and what depth of results you expect. The last few activities, however, are more suitable for upper elementary or junior high students.

Each activity is independent of the others—they could be done in any order. In each of the activities calling for geometric figures, the children make figures as shown in figure 1. You may want to show the children how to make geometric figures from the paper strips before beginning any specific activity that calls for this procedure.

The materials that are needed for each activity are listed. The type of paper used can be changed, but the weight of the paper should be about the same. Some wallpaper is a good substitute for construction paper and old manila folders may replace the lightweight tagboard. A paper cutter makes speedy work of cutting the strips. You will need a lot of strips if you do many of these activities with a class. Masking tape is probably the best tape to use; a quick way to distribute masking tape to a large group of children is to put (or have each child put) a piece on the back of a plastic ruler.

Triangle Strips

Need: Strips of construction paper (about 2 cm by 30 cm)

Tape
Centimeter ruler
Scissors

In this activity the children measure and fold strips to see if certain given lengths make triangles and, if so, to see what the triangles look like. For example, given the lengths 8 cm, 8 cm, and 6 cm, a child marks these lengths on a strip, cuts off the extra, folds on the marks, and tapes the ends together to form the figure. (Fig. 1) Younger children will discover that some combinations of lengths will make triangles and other combinations will not, while older children may be able to see why the sum of the lengths of any two sides

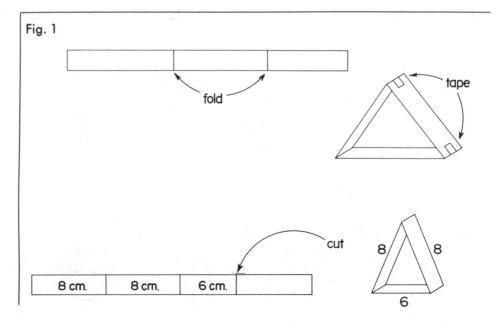

Fig. 1

fold

tape

cut

| 8 cm. | 8 cm. | 6 cm. | |

8   8

6

of a triangle must be longer than the length of the third side. You could begin with the following combinations of lengths:

(a) 10 cm, 10 cm, 10 cm
(b) 10 cm, 8 cm, 3 cm
(c) 10 cm, 10 cm, 8 cm
(d) 10 cm, 5 cm, 7 cm
(e) 10 cm, 7 cm, 5 cm
(f) 10 cm, 10 cm, 4 cm
(g) 10 cm, 10 cm, 1 cm
(h) 10 cm, 8 cm, 6 cm
(i) 10 cm, 8 cm, 2 cm
(j) 10 cm, 5 cm, 5 cm
(k) 10 cm, 3 cm, 6 cm

When some or all of these triangles (or other triangles) have been made, have the children see if any triangles are the same size (congruent) and which triangles look alike. (Lengths d and e make congruent triangles; i, j, and k will not make triangles.) Then have each child choose three lengths (the sum should be less than 30 cm) and see if the three lengths he or she has chosen will make a triangle. Or, have children fold a strip to make a triangle and then measure the sides.

Younger children can use the triangles for artistic creations. Older children may enjoy investigating isosceles triangles with a constant perimeter. For example, tell children that the perimeter of a triangle is 24 cm and let them see how many different isosceles triangles they can make if the lengths of the sides must each be a whole number of centimeters. (Answers: 11 cm, 11 cm,

and 2 cm; 10 cm, 10 cm, and 4 cm; 9 cm, 9 cm, and 6 cm; 8 cm, 8 cm, and 8 cm; 7 cm, 7 cm, and 10 cm) Have them try the same thing with perimeters of 18 cm, 25 cm, and 29 cm. Can the children find any pattern for the length of the unequal side of the isosceles triangle for any given perimeter? Does the pattern differ if the perimeter is an even or odd number of centimeters? Can children discover how many isosceles triangles are possible given any perimeter? (Try perimeters of 20 cm, 21 cm, 22 cm, . . . , 30 cm and see the pattern that emerges.)

Quadrilateral Strips

Need: Strips of construction paper
           (about 2 cm by 30 cm)
        Tape
        Centimeter ruler
        Scissors

This activity is similar to "triangle strips," but the children investigate quadrilaterals instead of triangles. Children will see that the four-sided figures are not rigid. (This is in contrast to triangles, which are rigid figures.)

The following sets of lengths will produce a variety of quadrilaterals:

(a) 8 cm, 8 cm, 8 cm, 8 cm
(b) 8 cm, 6 cm, 8 cm, 6 cm
(c) 8 cm, 8 cm, 6 cm, 6 cm
(d) 8 cm, 6 cm, 6 cm, 8 cm
(e) 6 cm, 8 cm, 8 cm, 6 cm
(f) 6 cm, 8 cm, 8 cm, 8 cm
(g) 8 cm, 6 cm, 8 cm, 8 cm

(h) 5 cm, 6 cm, 7 cm, 8 cm
(i) 5 cm, 6 cm, 6 cm, 7 cm

After a set of quadrilaterals has been made, have the children tell how any two quadrilaterals are alike and how they are different. (Quadrilaterals c, d, and e can be made congruent.) One thing to focus on would be the presence of sides of equal length and where they are in relation to other sides. For example, contrast b and c. Both quadrilaterals have two pairs of congruent sides, but in b the congruent sides are opposite each other, and in c they are next to each other. How does this change the looks of the figure?

Because quadrilaterals are not rigid figures, they can change shape. Children should experiment with the various quadrilaterals to see what shapes they can make. Can they make a square from a? Is it always a square?

Also, let the children make up their own sets of measurements and then make the corresponding quadrilaterals. A challenge is to have the children, without measuring, fold a strip into a rectangle (if its sides are held at right angles) that is not a square.

Math-Art Strips

Need: Construction paper strips
           (about 2 cm wide and any
           length)
        Centimeter ruler
        Tape
        Scissors

The paper strips may be used to make many artistic creations. If the children have made triangles or quadrilaterals from the strips, let them put the figures together to make their own creations. Or, you may want to suggest using strips to make flowers, animals, or funny faces. (Fig. 2)

If you want a more structured project, or if you want the children to have experience in following specific directions, have them make a funny face creation. The directions given in figure 3 are for fourth-grade or older students. Or you can simplify these directions for younger children by telling them when they start that they are making a funny face. Then, for example, instead of direction 4 (in fig. 3), just tell them to use triangles c and d for ears, without being concerned

Fig. 2

about the exact placement. In fact, it will be interesting to see the variations younger children make, even when they are all following the same directions.

Children may enjoy making a simple figure and giving directions to a friend on how to make one. This is not a simple task, but it is good practice in giving and following directions.

Number Strips

Need: White or light colored paper strip (about 2 cm by 30 cm)

Have the children fold their strips into eighths and number the parts 1, 2, 3, . . . , 8 on both sides—the same number should be on the front and back of each part. The children can then be told to fold the strip, backwards or forwards on any crease, to show different numbers. For example, have the children fold the strip so that only two parts are showing and then read the numbers as a two-digit number. Then ask them to fold their strip to show each of the following numbers:

56; 26; 81; 63; 12, 13, . . . , 18

Next you may want to have the children experiment with three-digit numbers. What three-digit numbers can they make with the number strips and using the same basic rules? Can they make each of the following numbers?

346, 128, and 568

Also ask them what number is showing on the back of the strip if 568 is on the front (answers will vary slightly).

Challenge older children to show three parts, each of which has an even digit on it, or to fold the strip to show all four odd digits.

The number strip can also be used to practice facts. "Front-Back," for example, is a game for two that provides practice in the addition facts. Have one child make a recording sheet for the two players by folding a sheet into halves and labeling one half *front* and the other *back* (fig. 4).

The two players decide who is "front player" and who is "back player." One player then folds the number strip to show two numbers. The front player records and adds the two numbers showing on the front of the strip and the back player does the same for the numbers showing on the back. They

Fig. 3

1. Make and label (write the letter on the strip after the extra has been cut off) the following triangles:

A	8 cm,	8 cm,	8 cm
B	10 cm,	8 cm,	10 cm
C	3 cm,	4 cm,	5 cm
D	3 cm,	4 cm,	5 cm
E	2 cm,	2 cm,	2 cm
F	2 cm,	2 cm,	2 cm
G	3 cm,	3 cm,	3 cm

2. Place triangle *A* on the desk so that one corner points directly to you. That corner will be called the bottom corner and the side opposite it will be called the top side. (All the remaining directions are given in reference to triangle *A* placed in this position.)

3. Place triangle *B* so that its shortest side fits exactly on the top side of triangle *A*.

4. Place triangle *C* on the outside of triangle *A* so that the corner between *C*'s longest and shortest side touches the top right corner of *A*, and *C*'s longest side touches *A*'s side.

5. Place triangle *D* on the other side of triangle *A* so that the figure now is symmetric.

6. Place triangle *E* inside triangle *A* so that one of *E*'s corners is on the top side of *A* and is 2 cm from the right top corner of triangle *A*.

7. Place triangle *F* in the same way as *E* except it should be 2 cm from the left top corner of triangle *A*.

8. Place triangle *G* inside triangle *A* so that two of *G*'s corners are each 3 cm from the bottom corner of triangle *A*.

9. What is it?

Fig. 4

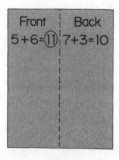

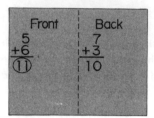

Fig. 5

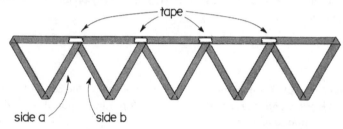

side a   side b

Fig. 6

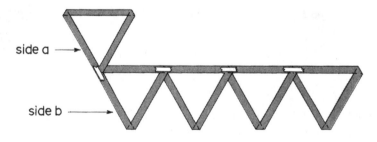

side a

side b

Fig. 7

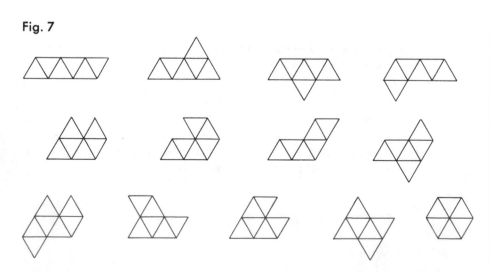

compare sums and the player with the larger sum circles it. Play then continues. Players take turns folding the strip into two parts—different parts each time. Players look only at their own side of the number strip as they fold it. They can play ten rounds and then see who has won the greater number of rounds.

Hinged Triangle Strips

Need: Tagboard strips (1 cm by 15 cm)
Tape

Have the children make an equilateral triangle out of each of five strips. They make the hinged triangle strip by taping the five triangles together so that the triangles can be folded easily back and forth at the hinges as shown in figures 5 and 6. After the triangles are taped across the top, each adjacent pair should be taped on the other side. This may be done easily by folding the pair so that two sides lie flat, such as shown in figure 6. Then have the children see which of the following five-triangle shapes they can make with their hinged triangle strips:

Next let the children try making the six-triangle shapes in figure 7 with their hinged triangle strips. This is done by surrounding a space with the sides of the triangles to make a sixth triangle. For example, begin with

and make *a* touch *b*.

In figure 7, the ones with a triangle like this

in them can be made.

GEOMETRY FOR GRADES K–6

Also let the children investigate what other shapes they can make with the hinged triangle strip. Can they make a shape with a pentagon in it? With a trapezoid in it? With a square or rhombus in it? What symmetric shapes can they make? Can they make the shapes shown in figure 8?

The hinged triangle strip may be varied by (1) using six triangles, (2) hinging the five triangles in different ways, such as

or (3) using squares instead of triangles.

Polygon Strips

Need: Lightweight tagboard or heavy construction paper
Strips (2 cm by 63 cm)
Millimeter ruler
Tape
Scissors

In this activity the children investigate two sets of polygons. Each set has polygons with three to eight, or possibly, nine or ten sides. In one set the perimeters of the polygons are the same, and in the other set the length of each side of each polygon is the same.

For the set of polygons with the equal perimeters have the children (alone or in groups) take six to eight strips. They fold one strip into thirds and make a triangle. Then, they fold another strip into fourths, keeping the angles as nearly the same size as possible. Another strip should be folded into fifths—it may be preferable to have the children first figure out how many millimeters long each fifth should be. From this strip they should make a pentagon, again keeping the angles as nearly the same size as possible. Have children continue with sixths, sevenths, and eighths) and ninths and tenths, if they wish). Once a set of polygons is complete, have the children stand each of the polygons on one of its sides and arrange the polygons in order—three sides, four sides, five sides, and so on—one behind the other. (Fig. 9) Then have the children look at the pattern of the heights of the polygons. What hap-

pens to the heights of the polygons as the number of sides increases? What happens to the areas of the interiors of the polygons as the number of sides increases?

For the set of polygons with sides of the same length, the children need six to eight strips. Have them make a triangle with each side 6 cm long, a square with each side 6 cm long, a pentagon with each side 6 cm long, and so forth. They should again try to keep the angles in each polygon as nearly congruent as possible. What happens to the heights of this set of polygons as the number of sides increases? What happens to the areas of the interior of the polygons.

Fig. 8

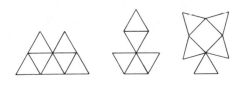

Fig. 9

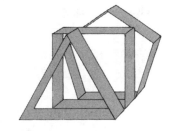

Fig. 10

Some children may enjoy making a mobile from each set of polygons.

Circle Strips

Need: Construction paper strips (about 2 cm by 30 cm)
Tape
Scissors

The children should have fun with and find some surprises in this activity, but the main purposes of the activity are to encourage children to make conjectures and to try variations on their own.

Have each child take two strips and make a circle with each strip. The two circles should then be securely taped together, on top of each other so that the planes of the circles are at right angles to each other. (Fig. 10) Be certain that the circles are taped well, otherwise the figure will fall apart when cut.

The children then cut each of the circle strips, one at a time, in the middle of the strip and completely around the circle strip. Cut through the tape, also. What is the result? (The children should get a square.)

See if the children can find out how to get a rectangle that is not a square. (They need to begin with two strips of different lengths.) Then see if they can make a parallelogram that is not a rectangle. (They need to place the two circle strips so that the planes of the circles are not at right angles.)

Have the children try putting the circle strips together in different ways and then have them predict what will happen when the circle strips are cut. Questions like the following will help the children start to try different possibilities, if they have not come up with their own. What happens if one circle strip is taped inside the other instead of on top of the other? What happens if three circle strips are used and taped together in different ways? With three circle strips, what about one circle strip inside the other and one on top? What about four circle strips? Is there a pattern based on whether the number of circle strips is even or odd? Let the children experiment with their own variations. As with all the strip activities, the children will come up with many surprising variations. □

TRACK CARDS

A Different Activity for Primary Grades

By **John A. Van de Walle**

Do the children in your class *talk* with their friends about mathematical ideas? Are they given the opportunity to plan, to try, to discover, and to create? The development of both the mental capability and the enthusiasm for creative expression and discovery must be nurtured at all age levels and all ability levels.

Track cards are a simple, easily made manipulative. Designs can be created, properties can be investigated, verbalization can be encouraged, and young children can have fun doing good mathematics. The materials described here were originally thought to be most applicable in the special education classroom to encourage verbalization and promote eye-hand skills. Indeed, in terms of spatial perception, eye-hand coordination, and simple problem solving, they have been well received by teachers of special students. However, while exploring the potential of the track cards, I have found them to have considerable value for all children in kindergarten through third grade. Even with an informal approach, the children become quickly engrossed in a variety of good problem-solving strategies. As the children construct roads according to a pattern, or create

An assistant professor of elementary education and mathematics at Virginia Commonwealth University in Richmond, John Van de Walle teaches content and methods courses at the graduate and undergraduate levels. He also supervises elementary student teachers.

a unique design, a situation is developed in which they (with or without the teacher) can think, discuss solutions, talk openly, and pursue creative ideas. These aspects alone make the track cards valuable in the primary classroom.

("Track Cards" is the name given to a similar set of materials that are a small portion of the kindergarten program of the Comprehensive School Mathematics Program, a division of CEMREL, Inc. In the CSMP materials, straights and curves are used to discuss the topological concepts of open and closed curves. All discussion

and extension of these materials in this article are the author's.)

The Cards

Track cards are made from tagboard or any lightweight cardboard cut into squares. The squares can be of most any size from about eight to twenty centimeters on a side. Larger, sturdier cards are suggested for the very young, but even 10-cm tagboard squares have been used with success in kindergarten. Each card has one of four types of "track"—"straight," "curve," "cross," and "tee"—drawn about 2 cm wide

Fig. 1

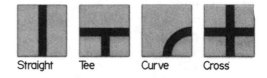

Straight Tee Curve Cross

Fig. 2

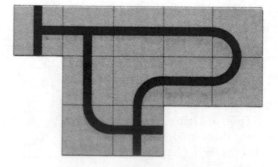

with a dark felt marker. (The cards could also be printed on an offset machine or the tracks could be cut from black paper and pasted on the cards.) A suitable collection for one to four children consists of about sixteen each of the "straight" and "curve" cards, and eight each of the "cross" and "tee" cards. (Fig. 1) The track on each card should meet the edge of the card exactly in the middle. Then any two cards placed edge to edge will form a continuous path. (Fig. 2) These paths (tracks, designs, patterns) are the objects of the various activities with the cards.

An optional layout board can be used to aid in the orientation of the cards. (Fig. 3) Such a board consists simply of a large grid with squares exactly the same size as the cards. A ten-by-ten square grid is adequate for most purposes, but the exact dimensions of the layout board are not important. I have found that a light-colored piece of flannel cloth stretched over a large piece of cardboard with the grid drawn on the flannel helps to keep the cards from sliding around as the children manipulate them.

Fig. 3

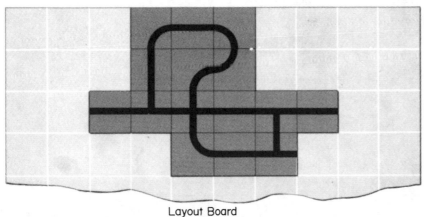

Layout Board

Fig. 4

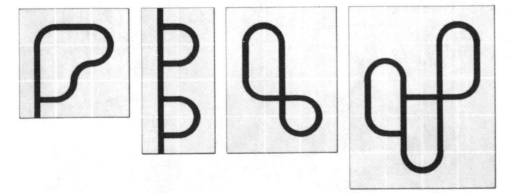

The Activities

1. Free play

As with all new materials, children should be permitted ample time to explore the cards in their own ways. One child or small groups of children will see quickly how the cards can be laid out to make designs, paths, patterns, and pictures. After only a few minutes of introduction to the cards, the children should be allowed to spend at least one period following their own inclinations with the cards. During this time of exploratory activity, be sure to listen to the chatter of the children and observe the patterns they make with the cards. Build their spontaneous vocabulary and their ideas into your first more structured activity.

2. "Make this pattern."

Make patterns that the children can reproduce with the cards. You can use a sheet of paper of notebook size (21.5 × 28 cm) with a 2-cm square grid. As an introduction to this procedure of following a pattern, kindergarten children benefit from the use of a much larger grid pattern drawn on the board or a large newsprint grid taped on the wall. With the large demonstration grid pattern, the children all work together with the teacher's directions. Once they can reproduce patterns from the large grids, children will be ready to work from smaller versions.

To make patterns, simply take a wide-tip felt marker and trace a heavy path around the grid. As your marker enters each new blank square, decide which of the four track cards you want to represent next and draw the line accordingly. Start with very simple patterns and put only one pattern on a sheet. The patterns in figure 4 are arranged in order of increasing complexity. Make plenty of patterns— it only takes a few minutes for each pattern. A nice collection of twenty patterns could easily be drawn up in a single evening. If patterns are slipped into acetate sheet protectors, they become a durable collection of activity materials.

3. Letters of the alphabet

Have children try to make their own initials. With some modifications, all the alphabet can be made using the track cards. The letters with slanted lines (*M, N, V, W, Z*) will require that you twist some of the cards (rather than lay them down edge to edge) or that these letters be made more "curvy." In any case, working toward these solutions provides excellent problem-solving and verbal experiences. Using the cards to form letters can be a fun way to discuss the shapes of newly learned letters in the kindergarten class.

4. "How many cards?"

A nice follow-up activity is to ask a series of "how-many" questions as children complete a pattern with cards. Such questions are especially helpful to children who are still developing number concepts and counting skills. The following are only suggestions of the type of questions that can be posed at this time: How many crosses (or tees, straights, curves) in your pattern? How many dead ends? How many cards did you use? How many closed loops do you have?

In a similar manner, questions asking for comparisons can be posed and are very important to the development of numeric concepts: Which kind of card was used most (or least)? Are there more cards in this row or this one? Which loop do you think is larger and why?

(The question of row length may provide interesting insights into the child's ability to conserve length or number as discussed by Piaget. Consider rows *A* and *B* in each of the drawings in figure 5. For many young children the answer to the question of which row is longer may not be the one we would expect. The track cards provide several opportunities to conduct conservation task experiments informally.)

5. Charts and graphs

The how-many and comparison questions give rise very naturally to simple charts and graphs that children can make. As a start, help children construct a simple bar graph showing the number of each type of card used in their designs or track pictures. A 5-cm square of construction paper can be pasted in the appropriate row for each card in the design. Or, cubical blocks can be stacked up—making a tower of blocks for each card type, using one block per card. Some children will be able to make graphs easily. Children who have trouble making graphs may be helped by having a square of paper (or block) placed on each card of the design. Then, as they pick up each square of paper or block, they paste or put it on the graph. (Fig. 6)

After children have had some experience with simple graphs, they can make and compare charts or graphs of several different designs. With more experience, children may be able to simply color in squares on grid paper, instead of cutting and pasting squares of paper. The charts and graphs in figure 7 are two different ways to record data for the purpose of organization, comparison, and observation.

Be sure to ask the how-many and comparison questions while you are discussing the graphs and charts. In one kindergarten class, the how-many questions were asked after a simple graph had been built with cubes. Interestingly, the children referred to the graph instead of the track picture for their answers. The superiority of

Fig. 5

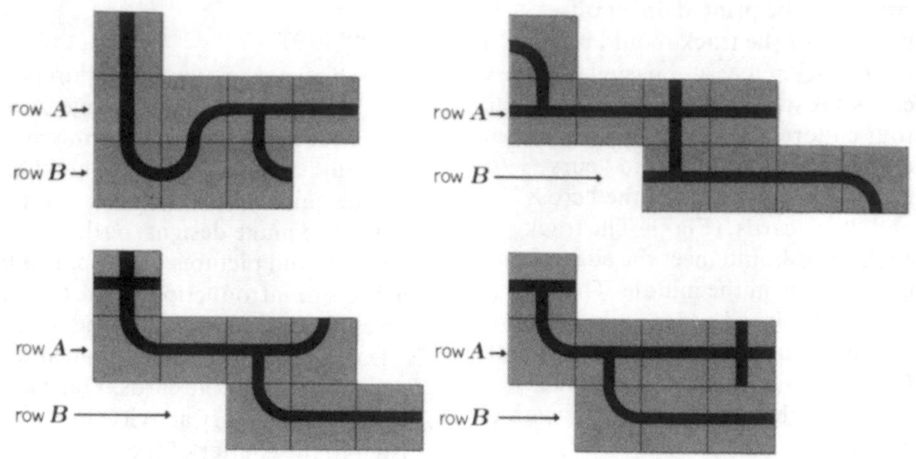

Fig. 6

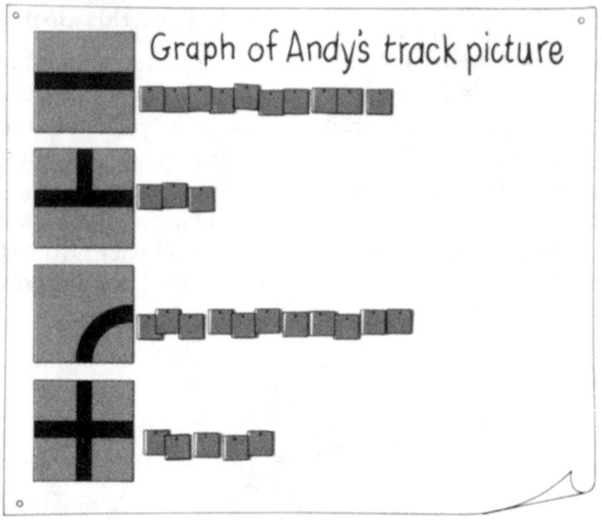

Graph of Andy's track picture

Fig. 7

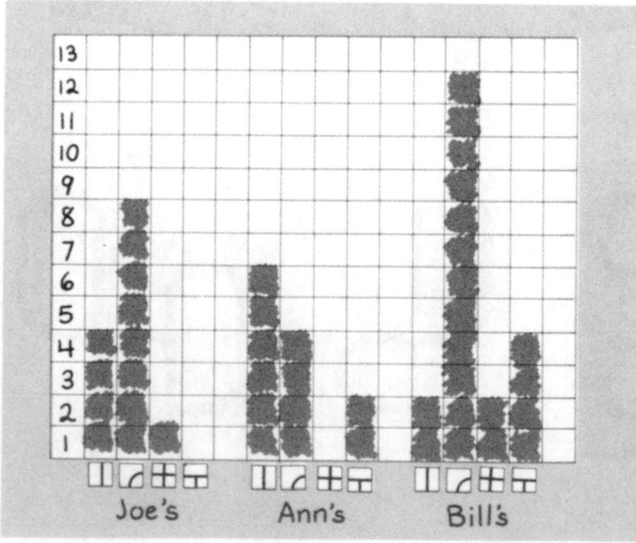

	Joe's	Ann's	Bill's
▯	4	6	2
◠	8	4	12
▯⊦	0	2	4
✚	1	0	2
Loops	2	1	3
Dead Ends	0	2	1
Total Cards	13	12	20

structured data and its use had been made evident, even to five-year-old children.

The representation of ideas is an extremely important aspect of mathematics that is too often overlooked in the primary grades. Young children find these track-card activities fun and exciting. Older children who have not been provided with such early experiences often struggle needlessly through text exercises involving graphical interpretations. The activities described here give children the opportunity to display work and create organized data from their own experiences.

6. "Make one like this."

All the activities mentioned in this article are suitable for children in kindergarten through second grade. This activity and those that follow may be slightly harder, but they can still be tried in grades one and two.

As a variation of the "make this pattern" activity, use patterns where no grid is marked and no indication of the individual cards is shown. (Fig. 8) To make patterns like this, place a grid sheet under a blank sheet of paper so that the grid shows through. Then draw the design as in Make This Pattern. When children are working from patterns without grids, it is important that you not be concerned

with exact duplications of these designs. Many of the children's designs will be somewhat different from what you expected, but you may wish to say such things as, "This design has two loops—what about yours?" or "It looks like you're going to need a tee card pretty soon." In this way, the essentials of the design (topological properties) are emphasized. Two children may make two different designs from the same basic pattern. When this happens, an interesting discussion can develop around how the two designs are alike and how they are different, and why they are both like the original pattern.

7. Create and record

In activities 2 and 6, the children lay the cards out according to a plan or pattern. There is considerable value in reversing this procedure. That is, when children have created a design on their own with the cards, they should have the opportunity to make a picture of their work. This can be done on either grid paper or plain paper (each requires different skills) with crayons or felt markers. Children also can be encouraged to write a few observations about their patterns to go with their drawings. Their observations may include such comments as "my pattern has a loop inside of another loop," or "this pattern has two sides the same,

you can fold it in two equal halves." The observations will range from the trivial to the sophisticated and will often be very nonmathematical. (Can we really complain if we have included a bit of language and writing practice in a mathematics activity?) Encourage children to make a little booklet of their designs and observations, or have a group of children display their ideas on the bulletin board.

8. Investigating symmetry

Track cards provide an interesting way to develop the concepts of symmetry. Start with the idea of line symmetry. With track cards, lay out a simple pattern that has a definite line down one side. It will be easier to talk about this line if the layout board is used. (Fig. 9a) Explain that you have made one half of the design and the children are to make the other half so that it looks like your half but is on the other side of the line. It may be necessary to discuss symmetry or mirror image, or even to complete one design to convey to the children what is to be done. Show them other examples of symmetry, such as leaves, butterflies, or folded paint blobs.

Another way of introducing line symmetry is to draw the first half of a design on grid paper, indicating the line of symmetry. Using this pattern, the child makes the pattern with the cards and then completes the design, creating a mirror image. If a mirror is placed on the line of symmetry in the pattern, the result should look like the completed symmetrical design. (Fig. 9b)

Once the concept of line symmetry has been established with the track cards, a number of variations provide interesting and challenging follow-ups. Activities like the following can be used.

(*a*) What happens if the line of symmetry does not touch the design? (Fig. 10)

(*b*) Try to construct a figure *on your own* that has a line of symmetry or "center line." Copy the design you make on a piece of grid paper and use a mirror to check the symmetry.

(*c*) Can you make a design that has two lines of symmetry, one vertical and one horizontal? (Hint: Draw the two lines of symmetry and start with a

Fig. 8

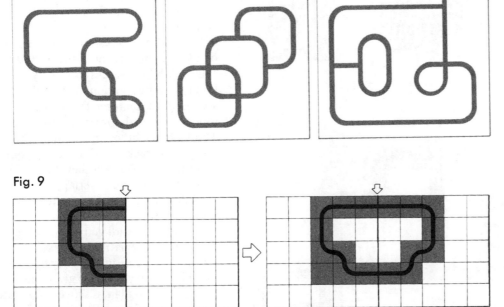

Fig. 9

design in just one corner. Now work from there.)

(d) Make a small design. Make its mirror image. Now make a mirror image of the mirror image. What happens?

9. Stretch a design

Another interesting activity is to distort a design in one or two directions—that is, "stretch it" up and down or sideways, or both. The number of loops and intersections does not change when the design is stretched. (These are known as topological properties of a curve.)

It is best to start with the unstretched design on a pattern card so that it can later be compared with the stretched version. (Fig. 11) It will require a bit of teacher assistance to convey to children the idea of stretching a design, but once a child has learned to distort a picture, many open-ended activities and opportunities for creativity will be provided.

Of course, there are no right answers to this activity. Your discussion should include such questions as, How are these still alike? How did you change it? What is different?

10. Verbal problems

One very valuable aspect of the track-card activities is that they provide a

setting in which kids tend to be verbal. Even young children can easily talk about the designs they are making and the problems they are working with. Within this free and informal context, you can introduce verbal problems, either written or oral. The purpose of the problem may be to capitalize on a certain concept (small-number counting, for example) or to develop verbal communication or listening skills. In any event, the track cards provide a chance to design real word problems in an easy, nonthreatening atmosphere. The following are just a sample of the problem-solving tasks you can try:

(a) Make a track with three loops.

(b) If you were building a road from point A to point B, what is the smallest number of track cards you would need? What is the smallest number of curves you could use? Could you do it with all curves? If so, what is the minimum number of curves?

(c) Make a track so that it has two loops. One of the loops should go around this doll and the other around this toy car but the loops should be connected. See if you can do it so that both loops have the same size (or shape or number of cards).

(d) Today you are a road engineer, but the people you work for have only

given you four curves and five straights. Can you build a loop road? (This one is impossible. Four straights is much easier, but think of the dialogue that this might provoke.)

(e) Can you build a path that has two crosses and two loops? What is the fewest number of cards you could use to do it?

(f) Try to build a track design that looks like a school bus (or a dog or a house).

(g) Make a path that is all connected but has one loop inside of the other, and two ways to get from the outside loop to the inside loop.

(h) Return to the charts and graphs discussed in 5. Use these in reverse to create problems. For example, road-builder McCoy wants a road built. He gave us this chart to go by. We must build the road the way he wants it, and the chart tells us what the road must have. Can you help him?

Summary

The activities presented in this article are intended to be only the beginnings of ideas rather than rigid prescriptions. Teachers and children alike will certainly invent new activities and create variations of other ones.

But first—make some cards! Just as an experiment to get started, cut up one or two sheets of tagboard and make a personal set. *Try* these activities *yourself*, at home or in the teachers' lounge. Then begin to think how your students could benefit from these activities.

It would be interesting to make a list of all the concepts and skills (including art and language areas) normally taught to kindergarten, first- and second-grade children that can be found in the activities described here. Think about this as you wonder how you can fit track cards in with all your other lessons. Perhaps they don't just fit in—perhaps they overlap.

Children (and teachers) who do not (or do not try to) think beyond "getting the answer" are not getting out of the world of mathematics those skills that are most valuable in everyday life. Are the children in your class given the opportunity to plan, discover, or create mathematical ideas? Perhaps track cards will be a beginning.□

Fig. 10

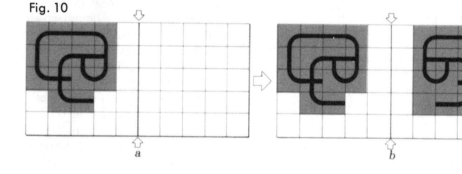

a

b

Fig. 11

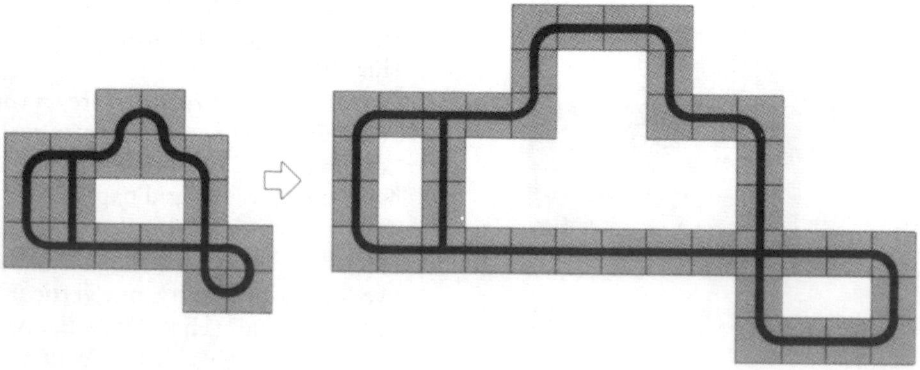

Frame Geometry: An Example in Posing and Solving Problems

By **Marion Walter**

"Close your eyes and try to see your favorite picture. Then open them and draw it." Children who do this will usually enclose their pictures in a rectangle or frame and often take great pride in the details of their drawings (fig. 1).

After talking about the drawings, you could discuss how picture frames look and how they are made. We will ignore, however, the width of the frames and problems like mitering the corners—at least at the beginning. Imagine that the frames are made from precut, thin sticks, which the frame maker obtains from a stick factory. Usually the sticks can be ordered precut in any length but, alas, at the moment the factory can supply sticks of only three different lengths: 3 cm, 5

Marion Walter is an associate professor in the Department of Mathematics at the University of Oregon in Eugene, where she teaches mathematics to prospective elementary school teachers. Her interest and work in informal geometry dates back to the early 1960s. She is the author of Boxes, Squares and Other Things, *an NCTM publication.*

cm, and 8 cm. (Obviously the frames are for miniature pictures, perhaps model houses, stages, or doll houses.)

One Starting Problem

Assume that the frame maker has as many sticks of lengths 3 cm, 5 cm, and 8 cm as we want. How many different shapes of frames can be made?

I purposely do not explain what is meant by *different* at the beginning, but the students soon ask. Is a frame that is 3 cm by 5 cm the same as one that is 5 cm by 3 cm? Students may, of course, pose this question just by pointing to two pictures (fig. 2, *a* and *b*) and asking, Are they the same? What will you and they decide? Is the hook already at the back of the picture frame? Someone may suggest that two hooks be put on the back of the picture so that it can be hung both ways. Such ideas can lead to fruitful discussions and further problems. For now we will decide that the two frames in figure 2 will be considered the same.

Try to solve the problem yourself, before reading on.

How do children solve the problem?

You may wish to supply the children with precut sticks. Older children could measure and cut the sticks themselves. You can use wooden sticks and plastic joints, or straws and pipecleaner joints. I have also used very thin strips of card and glue, and even chains of large paper clips (fig. 3). In any case, one way of solving the problem is to make all the possible frames of some material. Some children do this in a haphazard way and perhaps do not obtain all the possible frames; other children may use a system to obtain them all. One bonus that the children who use the sticks or straws method reap is that they are surprised to find that the model is not rigid!

On another level, children may try to draw all possible frames and using grid paper helps them. Again, some children draw the different frames in random order, while others try some system.

More advanced children may just write out the dimensions of each frame. They, also, may not find all the frames at first trial. Figure 4a shows a list of a student who used no system, and 4b and 4c show lists of two students who used a system. Can you see what systems were used in lists b and c? Note that in c the student listed all the squares first; then, all the non-square rectangles.

Some students try and solve the problem "just by thinking." They multiply three by three to get nine. Why does this not give the correct answer to "our" problem? To what problem is it the correct answer?

Children who have had some experience with problem solving and pattern examination sometimes will solve the problem by first considering the case of only one available length, and then two and three. They first may count all the squares and then all nonsquare frames to obtain a table. (See the first three lines in table 1.) Some children go on to find how many frames can be made if four lengths are available. They see that there are four squares, and that the fourth new length can be combined with each of the three old ones to give three new nonsquare frames. For five lengths we have five squares and four new nonsquare frames, giving a total of ten nonsquare frames and fifteen in all. Some youngsters go on to solve the problem for larger numbers of sticks and for the general case of n sticks. It is a nice problem that leads to experiences with triangular numbers. Many students notice how the numbers increase, as well as other patterns in the table. No doubt your students and you will find other ways to solve the problem.

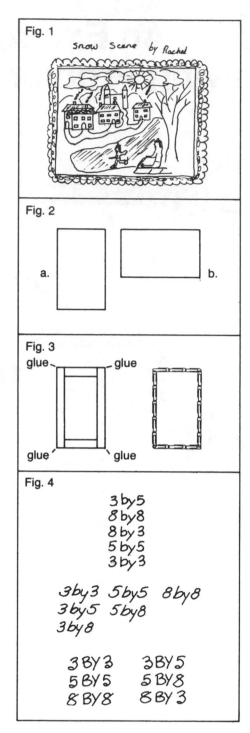

Fig. 1

Snow Scene by Rachel

Fig. 2

a.   b.

Fig. 3

glue — glue

glue — glue

Fig. 4

3 by 5
8 by 8
8 by 3
5 by 5
3 by 3

3by3 5by5 8by8
3by5 5by8
3by8

3 BY 3   3 BY 5
5 BY 5   5 BY 8
8 BY 8   8 BY 3

Some children will realize that you can find out how may frames are possible without knowing the actual *lengths* of the three sticks. That is, if you repeat the problem with any three different lengths—say 6 cm, 11 cm, and 19 cm—the answer will be the same. For such children, it may be a good time to introduce the notation a, b, c for the given lengths.

Variations

How many frames are possible if a 3-by-5 frame is considered different from a 5-by-3 frame? That is, we consider the two frames shown in figure 2 as different. The results are shown in table 2. Here we have the square numbers appearing!

Someone may suggest that each side of a frame can be made of two or more sticks. (You must decide whether you investigate or worry about whether two sticks would make a rigid side.) If you limit the conditions to two sticks per side, what lengths can be made now from 3-cm, 5-cm, and 8-cm sticks? Notice, for example, that although $3 + 3 = 6$ and $3 + 8 = 11$ give new lengths, $3 + 5 = 8$ does not. What numbers could you choose for starting lengths so that all sets of two sticks give a length different from 3, 5, and 8? If we choose 3, 5, and 7 for starting lengths, we get all lengths different from 3, 5, and 7, but $3 + 7$ and $5 + 5$ give the same length. What starting lengths could be combined in pairs to give all new *and* different lengths? Here we have a rich arithmetic problem—especially if you allow the combining of three lengths per side, or consider four starting stick lengths!

Table 1

| | Numbers of— | | |
different lengths	different squares	non-squares	Total
1	1	0	1
2	2	1	3
3	3	3	6
4	4	$3 + 3 = 6$	10
5	5	$6 + 4 = 10$	15
⋮	⋮	⋮	⋮
n	n	$\dfrac{(n-1)(n)}{2}$	$\dfrac{n(n+1)}{2}$

Table 2

| | Numbers of— |
lengths	frames
1	1
2	4
3	9
4	16
⋮	⋮
n	n^2

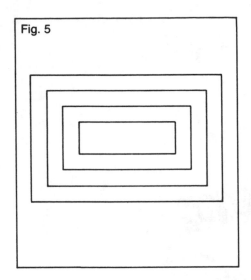

Fig. 5

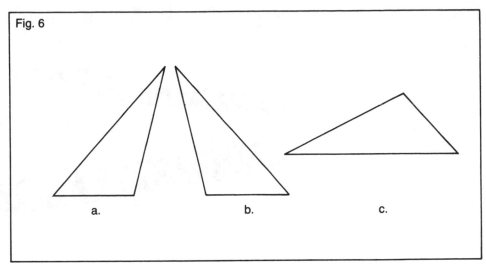

Fig. 6

a.　　b.　　c.

Another problem may arise. Note that a 3-by-8 frame has the same shape as a $(3 + 3)$-by-$(8 + 8)$ frame, and we can discuss frames that have different sizes but the same shape.

Still another problem that sometimes is suggested is to start off with a long stick of given length (say 20 cm) and ask how many ways can such sticks be cut into four pieces to make frames of different shapes and sizes (each side a whole number of centimeters). You or the children will have to decide whether the whole stick must be used.

If you consider the amount of paper inside the frame, the matting, or the size of the picture, you are led naturally into considering area as well as perimeter problems. Of course, you can repeat the problems with wide frames and consider inside and outside dimensions! (At a recent workshop, a teacher started exploring how such frames fit snugly inside each other—see fig. 5. Such nesting frames would store very economically.)

Just when you think you have done enough with the subject of frames, a student will suggest exploring triangular, trapezoidal or other polygonal frames. Before long, children are investigating which sets of three sticks can make a triangle—they are surprised that a 3-cm, a 5-cm, and an 8-cm stick do not. We had no such problems with rectangles! Who said that smaller numbers always involve easier problems! How many different triangles will sticks of length a, b, and c make if all combinations of the three chosen lengths make triangles? Children will have to decide whether the two triangles in figs. 6a and 6b are "the same." So here we will have to consider not only turns, as in figure 2a and 2b and in 6a and 6c, but also flips. And what if you considered only equilateral or isosceles triangles?

Conclusion

This article is only a brief outline of the paths that might be followed in working with one unit in informal geometry with children from a wide range of age and ability. The unit leaves plenty of scope for children to not only come up with new ideas, but to work through some of them. (For a way to extend the work into three dimensions, see Kuper and Walter 1976.)

The final "right" answer is not the only thing that matters. Posing and solving problems, clarifying problems, and finding different ways of solving the problems are also important activities. When children get different answers because they have interpreted a problem in different ways, rich and useful discussions can follow. In that way mathematics is seen as more than an endless series of worksheets, often done in isolation and checked routinely by the teacher or by students with an answer-book. Children see geometry as more than learning a bunch of definitions, identifying and classifying shapes, and applying formulas. To give just one example, in this unit a discussion about the difference between a square and a rectangle comes up naturally because that distinction is significant in the discussion; it is more than one, narrow objective— "to learn what a square is"—on a page. Note also that the arithmetic problems emerge freely from some geometric notion and need not be isolated into different chapters.

You may find it worthwhile to jot down some of the many mathematical concepts involved in working with this unit. Your list may make you mistrust the narrow behavioral objectives given on each page of so many textbooks and may make you question whether the structures of such texts are not restricting the content and activity in informal geometry in elementary and junior high school.

Reference

Kuper, M., and M. Walter. "From Edges to Solids." *Mathematics Teaching* 74 (March 1976): 20–23. ◗

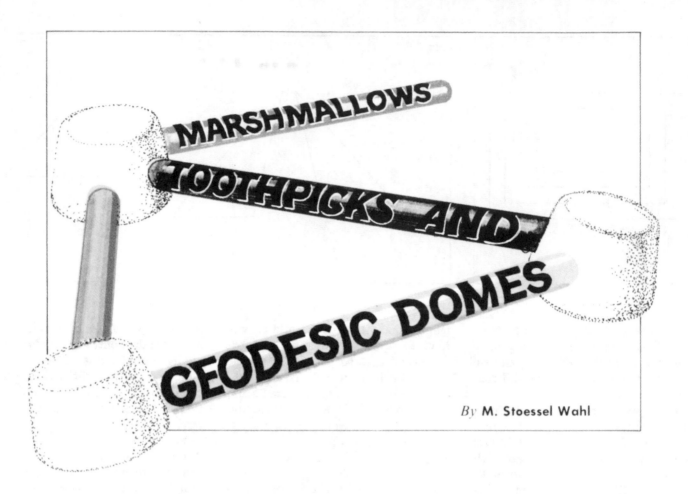

By M. Stoessel Wahl

Young children grow up in and adapt to a world of three-dimensional objects. Solids, however, are difficult to show on textbook pages and in chalkboard drawings, with the result that there is a tendency to limit early school geometric experiences to recognition of plane figures.

One successful learning approach to the three-dimensional world for young children can be made by having children construct geodesic domes with marshmallows and toothpicks. Since it involves physical activity, this approach also has been successful with young hyperkinetic children. The classroom construction of geodesic domes is an activity that is relevant to the world outside school, interesting for children, and fairly inexpensive to implement. Using miniature marshmallows and

An associate professor of mathematics at Western Connecticut State College, Stacey Wahl has a special interest in the creative teaching of mathematics. Much of her material is first tried out in the classrooms of former students. She and her husband are the authors of I Can Count the Petals of a Flower, *a counting book for young children, published by the NCTM.*

round toothpicks along with a well-constructed pattern, first graders can make a small dome. Larger domes can be made by children in higher grades.

Fig. 1

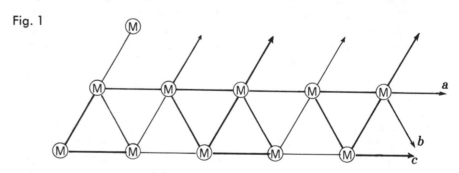

Basic pattern, 1 *v*, 5/8 geodesic dome

Fig. 2

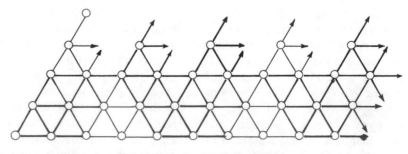

Basic pattern, 2 *v*, 5/8 geodesic dome

Patterns for Domes

Patterns for making two domes are shown in figures 1 and 2. The patterns

Using midget marshmallows and round toothpicks along with a well-constructed pattern, first graders can make a small dome.

are color coded to toothpick colors. (Colored cocktail toothpicks are available in supermarkets in four assorted colors—blue, green, yellow, and red.) When the larger domes are made cooperatively by several children, it is easier to assign sections to individual children by color. Even if plain toothpicks are used, it is helpful to have the pattern color coded. If the color coding is used consistently in all patterns, the various domes can be more easily compared and studied later.

Young children need patterns that are drawn to actual measures. To make such a pattern, first construct a dome. Then disconnect the dome from top to bottom and lay it flat on a piece of paper or cloth. Locate the positions of the marshmallows with black circles, and then draw in the colored connecting line segments so each pattern is in the desired one-to-one toothpick unit scale.

The storage of large permanent patterns often becomes a problem for a teacher. To minimize the storage problem, draw the patterns on light-colored cloth with dime-store crayons. After the pattern has been carefully checked

for accuracy, press it with a warm iron. The pressing melts the crayon marks into the cloth and the patterns can then be folded and put away—and even washed, if necessary.

Making Small Domes

Children in the primary grades can make the small domes (fig. 1). Give each child a teacher-made pattern (1 : 1 scale), eleven miniature marshallows, and twenty-five toothpicks. The child should spread out the pattern and then place the marshmallows, upright like a can, on each circle. The marshmallows are then connected with toothpicks, as indicated in the pattern. The toothpicks marked by arrows are not connected to marshmallows while the figure is in a flat position. Check that there is a bottom row of eight connected triangles and two partly opened ones. Then help the child tip this row on end and connect the open toothpick *a* to the upper marshmallow and tooth-

picks *b* and *c* to the lower one. The result at that stage should look like a baby's playpen. Bring the remaining four toothpicks into the single marshmallow at the top. The small geodesic dome is now completed.

Making Larger Domes

Children from the fourth grade up can successfully make the larger domes from patterns (fig. 2). The larger domes are more interesting and easy to make cooperatively. The miniature marshmallows should be dried, at least an hour on a dry day, before they are used to make a dome.

Before children attempt the largest dome (4 *v*), they should have experience in constructing smaller domes. Using several pairs of hands and a knowledgeable leader, the large dome is formed like the smaller ones; the lower row of triangles is connected first, and then, carefully, layer by layer, the upper triangles are put together. It is important to put the large dome over a volleyball or large inflated balloon to dry.

Since trouble often develops if the toothpicks are not well seated in the marshmallows, it is advisable to insert the toothpicks uniformly all the way

The larger domes are more interesting and easy to make cooperatively.

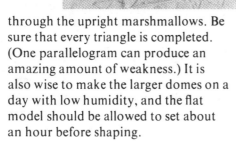

through the upright marshmallows. Be sure that every triangle is completed. (One parallelogram can produce an amazing amount of weakness.) It is also wise to make the larger domes on a day with low humidity, and the flat model should be allowed to set about an hour before shaping.

Damp weather does not affect gumdrops as readily as it does marshmallows, so some teachers might prefer to use gumdrops as connectors. The toothpicks, however, must be inserted at the correct mathematical angle in the gumdrops; the softness of the marshmallows lets the physical forces control the mathematical angle. If a more permanent, gumdrop model is desired, it would be wise to make a marshmallow model first; then make the gumdrop model one triangular layer at a time, carefully observing the angles of the toothpicks in each gumdrop.

Dome Sizes

The four sizes of domes are differentiated in *Dome Book II* by the designation of frequency: 1 *v*, 2 *v*, 3 *v*, 4 *v*, where *v* stands for frequency. The term *frequency* can be explained by reference to the construction of a simple icosahedron. If each side of each of the twenty equilateral triangles in the icosahedron is one toothpick in length, and if the lower five triangles of the icosahedron are removed, the resulting figure would be considered a 5/8, 1 *v*, geodesic dome. The dome is said to be of *one frequency* (1 *v*) because each of the basic equilateral triangles has sides one toothpick unit long. If each basic equilateral triangle had sides two toothpick units long, it would be a *two frequency* (2 *v*) model. Similarly, a 3-*v* dome would have basic triangles three toothpick units on each side; and the 4 *v* dome, four toothpick units on each side. The 4-*v* dome is quite large and requires skill, patience, and cooperation in its construction.

Figure 3 represents a schematic picture of a basic, triangular unit for each of the four different sizes of the domes.

Observations

The construction of domes of the different sizes can lead to other mathematical ideas and activities. One fourth-grade class placed the four finished domes one inside the other. They

Fig. 3

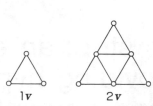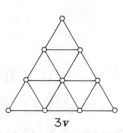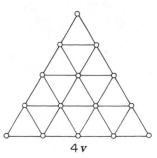

1*v*   2*v*   3*v*   4*v*

Basic triangles for domes of increasing frequency

were impressed by the changes in size and asked questions about area and volume changes. They had had previous experience with linear graphing and were quite surprised when their surface area graph produced a curved line. To graph the relative volumes they decided to line their domes with thin dry-cleaning bags and to fill them with big marshmallows. Thus their volume comparisons were made in big marshmallow units and they had a meaningful experience in making their first cubic graph.

Reference

Kahn, Lloyd. *Dome Book II.* Berkeley, Calif.: Pacific Domes, 1971. ☐

The Möbius strip: an elementary exercise providing hypotheses formation and perceptual proof

LLOYD I. RICHARDSON, Jr.

An assistant professor of elementary education at the University of Missouri—St. Louis, Lloyd Richardson teaches mathematics methods courses for elementary education majors. Previously he has taught mathematics content courses, directed student teachers, and taught elementary school grades 4 through 6.

An invitation to speak before the fourth-grade class of one of my former students had been accepted. As an enrichment lesson, I had decided to use the Möbius strip and to relate William H. Upson's story (Fadiman 1962), "Paul Bunyan versus the Conveyor Belt," to the class. Beyond telling the story, any attempts toward hypothesis testing or looking for patterns would depend on class reactions.

What is a Möbius strip?

A model of a Möbius strip can be constructed by putting a twist in a strip of paper and then attaching the ends. (See fig. 1.) A suitable material to use in making the strip is adding-machine tape. Using masking tape to attach the ends of the strip makes working with the twisted paper easier. To put an appropriate twist in the paper, rotate the end of the paper through 180°, then tape the ends of the strip together. Although some people consider a 180° rotation as a "half-twist," I label a rotation of 180° as a full twist. Children do not question or seem confused by this definition, but often they do seem concerned when the term "half-twist" is used—as though someone is doing only half of what he should.

Prelude to the Paul Bunyan story

Prior to telling the story, a few minutes were spent discussing with the children a method for determining the number of sides (or faces) on a sheet of paper. Usually, someone suggests marking each face with a different colored mark, then counting the marks. This suggestion was followed by examining the sides on a sheet of paper (discussing with students that the thickness of an edge of the paper is *not* to be considered as a side). It was a simple task to have a child shade the paper and

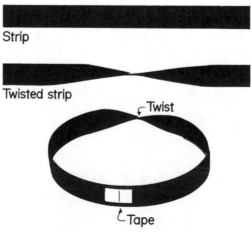

Fig. 1 Making a Möbius strip

show the class that a sheet of paper can be considered to have two sides.

A Möbius strip has one side

After the children and I had discussed the number of sides to a sheet of paper, it was rather easy for them to investigate the number of sides to a Möbius strip. Using a piece of adding-machine tape and masking tape, I showed the class how to construct a Möbius strip. Then each child was provided a Möbius strip and allowed to shade the strip by himself. Fourth graders seemed to have little difficulty in determining that the strip has only one side; though each child checked with his neighbor before volunteering his findings to the group. As one little boy said, "I see it and I know it's true, but . . . I don't believe it!" Now the children were ready to hear the Paul Bunyan story.

According to the story, Paul Bunyan ran a uranium mine in Colorado and used a conveyor belt to bring ore from the bottom of the mine to the surface. As the mine shaft was dug deeper into the ground, it became necessary to lengthen the conveyor belt. The conveyor belt was in the form of a Möbius strip, so Paul decided to cut the belt in two lengthwise. Charlie (Loud Mouth) Johnson bet a thousand dollars that if the belt was cut in two lengthwise, two pieces would result with each piece the same length as the original piece. Accepting the challenge, Paul cut the belt and the result was one strip, twice as long as the original piece. Alas, Loud Mouth Johnson lost one thousand dollars!

Six months later Paul told his foreman to cut the belt again as two narrower belts were needed to transport the ore to the surface. The ore had become smaller as the miners dug deeper into the earth, so a wide conveyor belt was no longer needed. Loud Mouth Johnson saw the foreman about to cut the belt and inquired as to what was happening. As Loud Mouth heard the foreman's description of events to come, he realized the foreman was taking the position that he had just lost a thousand dollars

in defending. So Loud Mouth bet the foreman a thousand dollars that cutting the belt would result in a belt half as wide and twice as long, the exact bet Paul Bunyan had used to win earlier. As you might expect, the cut resulted in two pieces and Loud Mouth lost a thousand dollars again!

The fourth-grade class remained quite silent as my scissors sliced lengthwise down the center of the Möbius strip, while I related the story of Paul Bunyan and the conveyor belt. As I finished the first cut and the strip unfolded to reveal one strip half as wide and twice as long as the original strip, the children raised right up out of their chairs. I had their attention for the rest of the period!

Once the story was finished, each child was allowed to cut a Möbius strip (the same strip he had shaded earlier). The children seemed amazed at the properties of the Möbius strip and caught on quickly to cutting and making Möbius strips.

HYPOTHESIZING

Telling the story of Paul Bunyan and allowing the children to cut a Möbius strip aroused much curiosity; everyone seemed interested in investigating the matter further. Since the twist we put in the paper seemed to play an important role, we decided to cut (lengthwise) strips with different numbers of twists and to keep a chart of the results of the cuts. Table 1 contains the results of our tests.

When a strip with zero twists was cut, the result was two pieces of the same length as the original piece. When a strip with one twist (a Möbius strip) was cut, the result

Table 1
Results of cutting strips with different numbers of twists

Number of Twists	Results
0	2 pieces
1	1 piece, twice as long
2	2 pieces
3	1 piece, twice as long
4	2 pieces

was a single piece twice as long as the original piece. Notice at this point that the children had been observing and recording data. After obtaining the data in table 1, the children were asked if they saw a pattern in the charted data. One child, as usually happens, suggested that for the even number of twists, two pieces result after cutting. We made a hypothesis at this point; the children were guided to observe what resulted when we had an odd number of twists. Again we made a hypothesis!

PROOF

The children were excited by their hypotheses and anxious to investigate them. We decided to try three test cases with an even number of twists (6, 10, 12) and three cases with an odd number of twists (5, 9, 11). Each strip was cut down the center lengthwise and the results investigated.

With the large number of twists in each piece, it was difficult to decide if one or two pieces had resulted from the cut. Thus we returned to the simple Möbius strip and two interlocking strips that had resulted earlier from cutting the strip containing two twists. It was decided that if a strip was torn in two, then one long strip would result from a single strip; but from two interlocked pieces, a single strip and an intact strip would result from tearing the strip. (See fig. 2.) We returned to the strips we had cut to test the hypotheses and carefully tore the strip. In each case the results confirmed the hypothesis.

DISCUSSION

The beauty of using the twisted strips was that they allowed a perceptual method of proving or disproving a hypothesis prior to a formal course or discourse in geometry. The method cannot exhaust all of the cases but children are provided an interesting and respectable approach to observing, hypothesizing, and validating a hypothesis. The strips provide a concrete approach to

informal proof. Frankly, I was amazed at the demonstrated thinking (through verbalization) that the period provided.

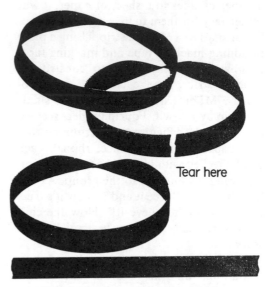

Fig. 2 Investigating the number of strips

The use of the Möbius strip and the story of Paul Bunyan has yet to fail to stimulate interest in other mathematics classes I have visited. Students generally become excited as they make the table showing the relationship of twists to results. Hypothesizing from the table generally is spontaneous within the class. Testing the hypothesis, although not mathematically rigorous, is definitely worthwhile in developing the notion of proof for the child prior to a formal introduction to proof.

Lastly, children are involved, excitement is in the air, and results are a confirmation or rejection of inductive logic—and that is the essence of mathematics.

Bibliography

Fadiman, Clifton. *The Mathematical Magpie.* New York: Simon and Schuster, 1962.

Mathematics for the Overhead Projector

By **Jill E. Arledge**

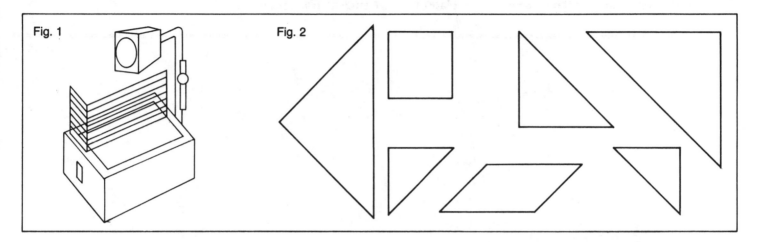

Fig. 1

Fig. 2

Mathematics can be interesting. As a matter of fact, mathematics can be presented in a way that makes it the most exciting time of a student's day. The key, I have found, is to get the student's attention. This can be done easily through the use of games.

Many games, however, are meant for only a few players. If you want to include the entire class on a given concept, then a way must be provided for doing so. The easiest way I have found is to use the overhead projector. My students love the idea and it is a good way to use a piece of equipment that is often not used. The overhead projector can be used in any subject matter area, of course, but I will describe here two games I have developed and used in the area of mathematics.

Jill Arledge is a fourth-grade teacher at St. Mark's School of Texas in Dallas, Texas. She has a masters degree in education from North Texas State University in Denton.

Symmetry

The general objective of this game is to promote a deeper understanding of symmetry. Visual perception is also developed.

To begin the game, two teams must be formed. Either one or two overhead projectors are needed. If only one projector is used, a divider that is about eight inches high must be placed down the center of the projecting plate so the two players cannot see what is being placed on the opposite side of the plate. (See fig. 1.) Two or more sets of geometric shapes are needed.

I begin by using sets of five rectangles of various sizes. The sets given to each team must be identical. The first player from team one arranges her or his shapes in any order while the projector light is off. Then the projector light is turned on and a player from team two must match the first team's design so that the two images on the screen are symmetrical to one another.

The team member trying to match the design is given a limited amount of time (two to five minutes) to get her or his game pieces in the right order. The added difficulty comes with the projection being inverted from the way the rectangular shapes are placed on the plate. If the first player cannot make the right arrangement in the time allowed, the next member of the team is given a chance, and so on, until someone gets the pattern. Each time a new member of the team is given a turn, the opposing team gets a point.

When the first design is made, the opposite team is given a chance to make a design. The game ends when everyone has had a try or when a given number of points has been accumulated.

After a few games are played, the excitement can be heightened by shortening the time limit and by using colored acetate for the game pieces. New shapes and additional pieces may be added as the students become more familiar with the game. Identical sets of shapes can also be distributed to the rest of the students in the classroom so that they can be working on the design while they wait for their turns.

Tangrams

The object of this game is to improve logical thinking through the use of tangrams. A previous knowledge of tangrams is needed.

Two teams of any number of students are needed for this game. Again, two projectors may be used, or the projector plate must be separated into two sections. It is better if two projectors are used because the images from the two machines can then be overlapped, making the task easier for the student.

When one projector is used, the images are side by side and not overlapped. At least two sets of seven-piece tangrams are needed. (See fig. 2.)

A student on team one places her or his set of tangrams on the projector plate in any way that forms a connected shape. The projector light should be turned off while the shape is being set. When the student is ready, the light is turned on and the shape is projected on the screen. Then on another projector (or on the other part of the plate) the first student from team two must try to match the first tangram shape. A time limit (three to five minutes) is set. If the first player cannot match the design in the time allowed, another member of the team tries to complete the task. For each new member that it takes to complete the design, a point is given to the opposing team.

For added group participation, I give each team member a set of tangrams so that they can have a chance to work on the design at their seats in anticipation of their turns. ◗

Teaching Geometry with Tangrams

By **Dorothy S. Russell** *and* **Elaine M. Bologna**

What is the most neglected area of the elementary school mathematics curriculum? The answer, probably, is geometry. Too many people think of geometry as a formal structure, like the course they had in high school. As a result, they do not see its relevance to the elementary school mathematics curriculum. Activities that introduce children to geometric concepts provide experiences that help children develop and reinforce spatial perceptions.

Tangrams can be used to provide numerous worthwhile mathematical experiences for children. As a teaching tool, they are doubly beneficial. They not only offer reluctant teachers a simple, but exciting, means of introducing geometric concepts, but also provide an excellent vehicle for students and teachers alike to engage in tasks that foster spatial visualization.

For those readers not familiar with tangrams, they are a seven-piece puzzle which reportedly originated in China. The seven pieces consist of two large right triangles, one medium-size right triangle, two small right triangles, one square, and on parallelogram. Countless designs can be formed from these seven shapes. Plastic commercial sets of tangrams can be purchased for less than one dollar

Dorothy Russell is director of teacher education and associate professor of education at Salem College in Winston-Salem, North Carolina, where she teaches courses in elementary mathematics education. Elaine Bologna is director of the Demonstration Mathematics Center/Summit School in Winston-Salem. She is a former member of the Editorial Panel of the Arithmetic Teacher.

per set. We recommend, however, that tangrams be introduced to children through the following paper-folding activity. (Several books offering additional activities involving tangrams have been listed in the references.)

Other than paper, pencil, and scissors, few materials are necessary to introduce students, through tangrams, to the following geometric ideas: triangle, square, trapezoid, parallelogram, right angle, hypotenuse, area, congruence, similarity, and symmetry. The Pythagorean Theorem may also be discovered. The authors have used these activities successfully with students in grades three through six. It is suggested, however, that students have some work with computing area of a square or rectangle before using the activity to discover the Pythagorean Theorem. Throughout the following activities, the emphasis is on student discovery. Readers will most easily follow the subsequent directions if they will *do* the constructions as they read through the article.

Constructing Tangrams through Paper Folding

Materials: sheet of paper (8 1/2″ by 11″), pencil, scissors.

(A set of the seven tangram pieces can be constructed from a transparency and then used on the overhead projector to demonstrate the various steps as you progress.)

1. Fold the piece of paper to make a square by bringing point *a* to point *b* and creasing the paper (fig. 1). Cut off the surplus. Discuss the shape you now have.

2. Now fold the paper again, along the other diagonal. Discuss the shapes (triangles) made by these folds. Cut along one diagonal to form two large triangles (fig. 2). Have children discuss the size of these two triangles.

3. Cut one large triangle along the center line to form two smaller triangles (fig. 3). Label them 1 and 2. Students can be asked questions like the following ones. (Related mathematical ideas are in parentheses.)

a. Are shapes 1 and 2 alike? How? (Congruency)

b. What do you notice about the corners of these triangles? (Size)

c. Are any of the triangles' corners square? (Right angle)

d. How can you prove corners are square? (Put the corner against the corners of floor tiles, a desk, cover of a book, room, and so on. Tell children that a triangle that has one square corner or right angle is called a right triangle.)

e. Are there any other square corners in the room?

f. Are all square corners congruent: (Have children prove this by putting a right angle up to other square corners or right angles. Introduce the measure of a right angle as 90°.)

g. How would you describe the sides of the right triangle? Are any sides longer than the others? (Introduce *hypotenuse*, the name of the side opposite the right angle.)

h. Look at the remaining triangle. Does it have a right angle? Where is its hypotenuse?

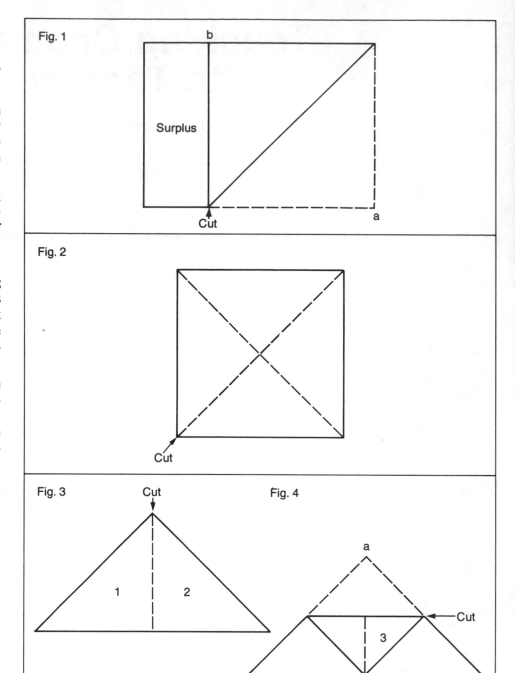

4. Holding the remaining large triangle, fold the corner or the right angle to the midpoint of the hypotenuse and crease (fig. 4). Cut along the crease and label the new triangle 3.

Questions for students:

a. Is shape 3 congruent to shapes 1 and 2? Does 3 have a right angle? Is it a right triangle? Where is the hypothenuse of triangle 3? What do you notice about the sizes of the other angles of the right triangles? (Shape 3 is similar to shapes 1 and

2. Each angle of one triangle is congruent to the corresponding angle of another of the right triangles.)

b. Ask students to describe the remaining piece. (A trapezoid is a four-sided figure with two parallel and two nonparallel sides.)

c. Does the trapezoid contain any right angles? Are any of its angles congruent to any other angles? How do you know this? (Place smaller angle of right triangle over smaller angle of trapezoid; place

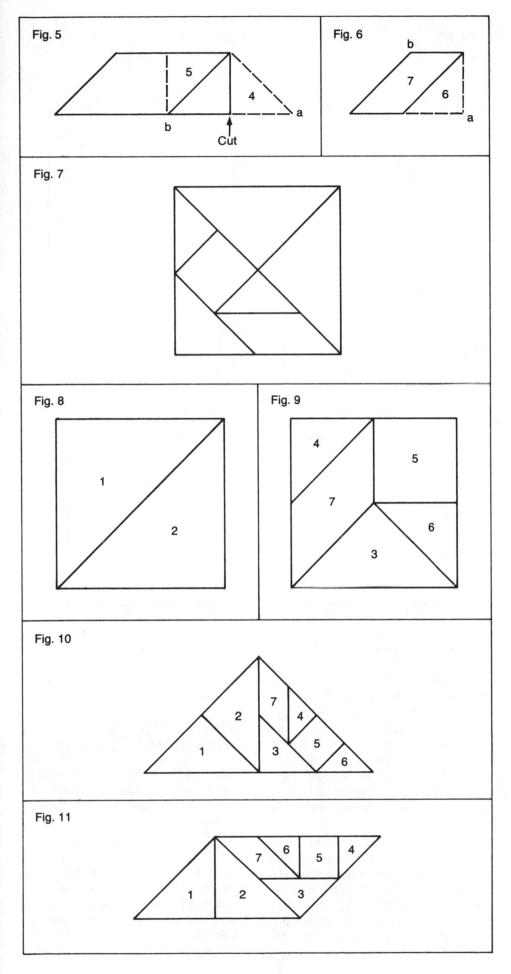

Fig. 5

Fig. 6

Fig. 7

Fig. 8

Fig. 9

Fig. 10

Fig. 11

right angle plus smaller angle over largest angle of trapezoid.)

5. Fold one corner, *a*, of trapezoid to midpoint of baseline, *b* (fig. 5). Crease. Cut along crease and label this triangle 4. Cut off remaining square and label it 5.

Questions for students:

a. Is 4 congruent to any other triangles? Does it contain a right angle? What kind of a triangle is it? Where is its hypothenuse?

b. Looking at 1, 3, and 4 how are they alike? How are they different? (Similar)

6. Fold the remaining piece so that you obtain another right triangle and a parallelogram (fig. 6). Bring point *a* to point *b* and crease. Cut and label the triangle 6 and the parallelogram 7.

Questions for students:

a. Is shape 6 congruent to any other piece? Is it similar to another piece? What do you know about the angles in 6?

b. How would you describe shape 7? (A parallelogram is a four-sided figure and the opposite sides are parallel.)

c. Does 7 contain any right angles? Are any angles of 7 congruent to any other angles of the triangles? (Help children discover that the smaller angles of the parallelogram are congruent to the smaller angles of the right triangles. Also, the larger angle of the parallelogram is congruent to a right angle plus the smaller angle of the triangle.

Spatial Visualization Activities

Materials: set of tangrams for each student or pair of students.

Before beginning, review the names and numbers of the seven pieces. How are the various pieces alike and different? The answers to the questions that follow are in parentheses.

1. How many other tangram pieces can you make with 4 and 6? (5, 7, and 3)

2. How many ways can you make shape number 1? (Three: 3, 4, and 6: 5, 4, and 6; 7, 4, and 6)

3. How many 4's (small triangles) are in 5? (Two) In 7? (Two) In 3? (Two) In 1? (Four) In 2? (Four) In 6? (One)

4. Put the seven tangram pieces back together again to make the square we started with. (*Hint:* Try to remember how you made 1 and 2. Where did 3 come from?) See figure 7 for the solution.

5. How many 4's make up the large square? (Sixteen)

6. Make a square using 1 and 2. How many 4's does this square take? (Eight) Using the remaining five pieces, try to make another square congruent with this square. How many 4's does the square you just made take? (Eight) See figures 8 and 9 to see how to make the squares.

7. Ask students to make a large right triangle or a parallelogram using all seven pieces. See figures 10 and 11 to check.

8. Have the children put all seven tangram pieces together to make a shape or design. Carefully trace the *outline* of this design. Paste the outline on a piece of construction paper and then laminate the construction paper. These can be used to make a collection of challenging puzzles for children. Have the children try to fill in the outlines with a set of tangram pieces.

9. Make some transparencies of the outlines of various shapes (figs. 12, 13, and 14).

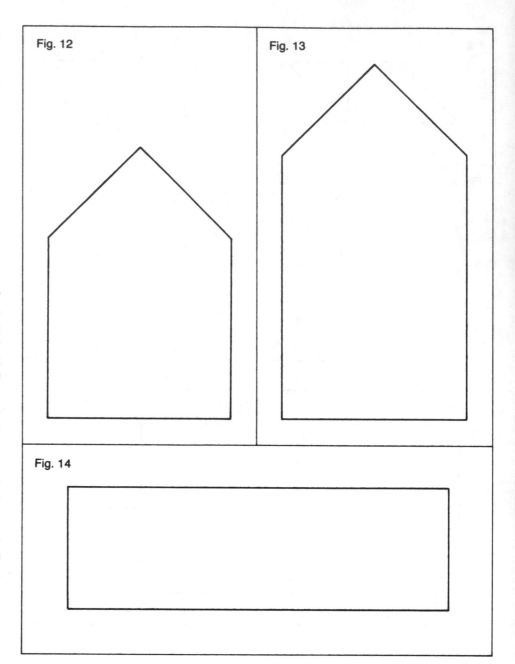

Fig. 12

Fig. 13

Fig. 14

Measurement Activities

Materials: set of tangrams for each student or pair of students.

The following are some activities that give children experiences in measuring angles.

1. Given that a right angle has a measure of 90°, what is the number of degrees in each of the other angles of triangles 1, 2, 3, and 4? (Demonstrate on the overhead projector with transparency pieces that each one has a measure of one-half of 90°, or 45°.)

2. What is a straight angle? (The two sides of a straight angle lie on a straight line.) How many degrees are there in a straight angle? Prove that the measure of a straight angle is 180° by placing two right angles together. (90° + 90° = 180°) How many right angles can you put around the center of a circle? (Four) How many degrees are on a circle? (4 × 90° = 360°).

3. Ask students to find the measures of each of the angles of the parallelogram. (The angles measure 45° and 135°.) The 45° angle may be discovered by placing the smaller angle of a right triangle over the smaller angles of the parallelogram. The 135° angle may be found by placing a right

angle and a 45° angle over the larger angle. (45° + 90° = 135°).

Discovering the Pythagorean Theorem

Materials: Pencil, several sheets of paper, set of tangrams (plastic or cardboard tangrams will make this activity easier).

Use the following procedure:

1. In the center of the paper, trace triangle 3. Label the hypotenuse, *c*, and the adjacent sides *a* and *b*. Mark the right angle with a small square at its vertex.

2. What two pieces can be put to-

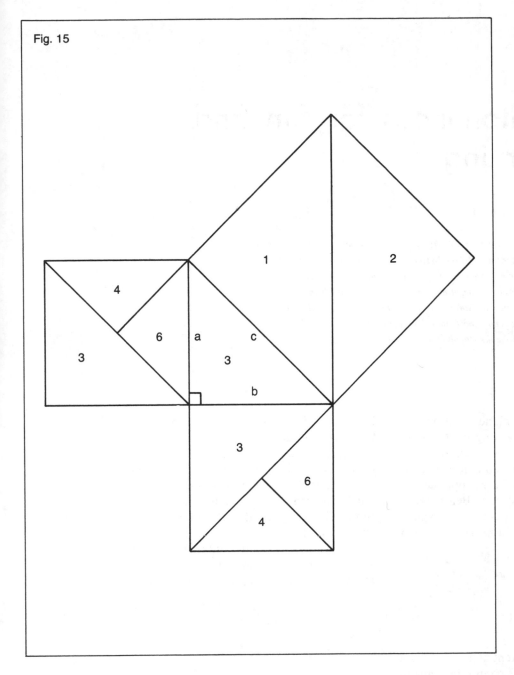

Fig. 15

References

Buell, Clayton E., Irvin Schwartz, and Alan Barson. *Activities with Tangrams*. Newton, Mass.: Selective Educational Equipment, Inc., 1978.

Fennema, Elizabeth and Julia A. Sherman. "Sexual Stereotyping and Mathematical Learning." In E. H. Werner, ed., *Sex Role Stereotyping in the Schools*. Washington, D.C.: National Education Association, 1980.

Jenkins, Lee and Peggy McLean. *It's a Tangram World*. San Leandro, Calif.: Educational Science Consultants.

Maccoby, Eleanor E. and Carol N. Jacklin. *Psychology of Sex Differences*. Palo Alto, Calif.: Stanford University, 1974.

Seymour, Dale. *Tangram Math*. Palo Alto: Creative Publications, 1971. ◗

make the squares on sides *a* and *b*? (1 and 2) How many 4's (small triangles) are in each square? (Square on side *c*, 16. Squares on sides *a* and *b*, 8.) Is there any relationship between the squares on the three sides?

7. What relationship holds true for the squares on the sides of each of these right triangles? (The area of the square on the hypotenuse is equal to the sum of the areas of the other two sides.) Is there any way you can express this relationship in a mathematical statement or formula? (Based on the students' prior experience in mathematics, they may say either of the following:

$$(c \times c) = (a \times a) + (b \times b) \text{ or}$$
$$c^2 = a^2 + b^2$$

The students have now discovered the Pythagorean Theorem.

NOTE TO THE READER: If funds are not available in your school to purchase a set of plastic tangrams for each student, you can make a useful set quite inexpensively. Use figure 7 as a model for a duplicating master, and duplicate a copy for each student. You can use construction paper by "hand feeding" the duplicating machine, or use duplicating paper. Laminate or cover each copy with clear Contact paper and then cut the pieces apart. Store each set in a ziploc bag. These make durable and reusable sets.

The authors have also made sets by tracing pieces onto heavy cardboard or mat stock, and cutting with a paper cutter.

gether to make a square on side *c*? (1 and 2) Trace the square as shown in figure 15.

3. Which tangram pieces can be put together to make a square on side *a*? (3, 4, and 6) Trace this square on side *a*. Repeat the same procedure for side *b*.

4. Ask the students how many 4's (small right triangles) are in each square. (The square on side *c* has 8; the square on side *a* has 4; the square on side *b* has 4.) Do you see a relationship between these three numbers? (The sum of the two smaller numbers is equal to the third number.)

5. Repeat steps 1–4, beginning with

the small triangle, 4. Trace 4 in the center of a piece of paper. Which tangram pieces will make the square on side *c*? (3, 4, and 6) Trace the square. Which tangram pieces will make the squares on sides *a* and *b*? (4 and 6) Trace these squares. How many 4's (small triangles) are in each of the three squares? (Square on side *c*, 4. Squares on sides *a* and *b*, each 2.) Is there any relationship between the squares on these sides?

6. Repeat steps 1–4 beginning with triangle 1. Trace triangle 1 in the center of the paper. Which tangram pieces will make the square on side *c*? (All seven pieces) Which pieces will

GEOMETRY IN UNUSUAL WAYS

Pentominoes for fun and learning

RICHARD A. COWAN

At the time this article was written, Richard Cowan was a mathematics field agent for the Del-Mod System at the University of Delaware, where he did inservice work with teachers, taught mathematics education courses, and was a resource consultant for the Science and Mathematics Resource Center at Delaware State College and Delaware Technical Community College. He is now coordinator for a mathematics basic skills project in Roanoke Rapids, North Carolina.

A pentomino is a plane figure that is formed by grouping five congruent, square shapes together so that every square has at least one of its sides in common with at least one other square. Two simple pentominoes are shown in figure 1. Before reading further, how many different pentominoes can you construct? Two

Fig. 1

pentominoes are different if they are not congruent; that is, if one cannot be made to match the other by flipping or by rotating.

In figure 2, for example, *a, b, c,* and *d* are the same pentomino.

A good classroom exercise is to have your students find and cut out of graph paper all of the pentominoes. This is a lesson in perception, originality, discovery, and double checking that the students will find enjoyable.

The author has found that for students in the lower grades, it is better if the students are provided with five individual square

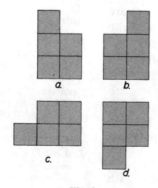

Fig. 2

Fig. 3

pieces of paper that are the same size as the squares on the graph paper. The children are then encouraged to construct the pentomino shape with the individual square pieces before they cut the pentomino out of the graph paper. In a one-hour period in one third-grade class in which children were doing this, two students found all of the pentominoes. There are twelve different pentominoes, as shown in figure 3.

Once all of the students have made a set of pentominoes, there are many things that can be done with them, depending on the level of the students and the desires of the teachers. Several examples of ways in which pentominoes can be used with classes are included here.

In grades 3, 4, and 5, as students find and cut out the pentominoes, the teacher can point out some of the geometric principles of congruency. For example, a teacher might ask the following questions:

Do all of the pentominoes have the same area? The same perimeter?

Some of the pentominoes can be folded up to make a cube without a lid. Which ones are they?

Using all twelve pentominoes, can you construct a 6-by-10 rectangle? (This is the puzzle used in the commercially produced game, Hexed. A solution is shown in figure 4.)

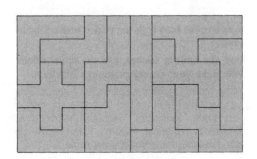

Fig. 4

Other problems that would be appropriate for older students include the following:

Can you find two different pentominoes that can be used to make a 2-by-5 rectangle?

Complete the following table.

Number of pentominoes	Size of rectangle	Can it be done?
N	N × 5	Yes or no
2	2 × 5	?
3	3 × 5	?
.		
.		
.		
12	12 × 5	?

Can you find three different pentominoes that can be used to make a 3-by-5 rectangle?

Given any pentomino, use nine of the other pentominoes to construct a scale model three times as wide and three times as long as the given piece.

There are many activities on tessellations or tilings with the pentominoes that could be used if a teacher finds tessellations interesting. Figure 5 shows two tessellations. By using the pieces for coverings, the pento-

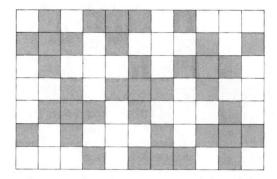

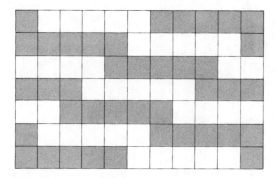

Fig. 5

minoes also can be used to develop young students' concepts of area.

There is also a game that can be played with the pentominoes and a checkerboard made with squares the same size as the squares in the pentominoes. The purpose of the game is for students to see if they can place all of the pentominoes on the checkerboard without overlapping the pieces. The game can be played in several ways. It can be a game of solitaire; or, two players with one set of twelve pentominoes can play against each other on the check-erboard. Each player draws six pieces from the set, and then, in turn, each player places a pentomino piece on the board until no more pieces can be played. The last person to place a piece on the board is the winner.

More than two also could play the game. In another variation, the person who in three games has placed the most pieces on the board is the winner.

These are only a few of the things that children can find to do with the pentominoes. You and your students may have fun finding others.

Geometry with a Mira

ERNEST WOODWARD

A professor of mathematics at Austin Peay State University,
Ernest Woodward teaches both mathematics and mathematics education courses.

Are you looking for a device that will help you teach the ideas of symmetry, congruence, and reflections; a device that will make the construction of parallels, perpendiculars, and bisectors easy and intuitive? If your answer is yes, you should consider the possibility of using a Mira. The main portion of the Mira is a piece of translucent red acrylic plastic about 9cm by 15cm. One of the 15-cm edges is rebated. The Mira is held upright by two ends, which can be made of either plastic or wood. The purpose of the ends is to make the device sit perpendicular to the surface being examined. When the Mira is used, the rebated edge is placed down and toward the user so that lines can be drawn as in figure 1.

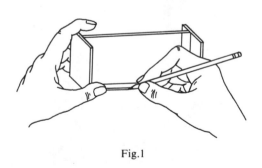

Fig.1

The most obvious use of the Mira is for line reflections. The Mira can be placed on the reflection line and the image can be sketched by drawing behind the Mira while looking through from the front side. This can be done quickly and easily and is particularly convenient for compositions of line reflections. Since the plastic is translucent, it is possible to see an object through the plastic in addition to seeing the image of anything on the forward side of the Mira, as shown in figure 2. This enables

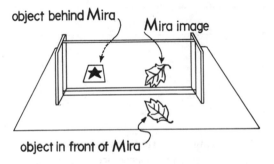

object behind Mira

Mira image

object in front of Mira

Fig. 2

a user to find symmetry lines for plane figures by merely adjusting the Mira so the image of the portion of the figure on the forward side of the Mira matches the portion of the figure on the reverse side of the Mira. When a single piece of paper is used on a hard surface, and when lines are drawn along the rebated edge, reasonably accurate symmetry lines result. Since the perpendicular bisector of a segment and the bisector of an angle are both symmetry lines, these constructions are trivial. With a little care, it is possible to demonstrate that for a triangle, the angle bisectors are concurrent and the perpendicular bisectors of the sides are concurrent.

A line perpendicular to a given line through a given point can be constructed readily. This is done by placing the Mira so that a portion of the line matches the image of the other portion of the line and then sliding the Mira (in this perpendicular position) until the desired point falls along the rebated edge. Using this technique, it is simple to show that the altitudes of a triangle are concurrent.

The Mira can also be used to construct a line parallel to a given line through a given point. To construct a line parallel to line l_1 and passing through A in figure 3, use the Mira to draw line l_2 perpendicular to l_1 and passing through A, and then draw line l_3 perpendicular to line l_2 and passing through A. Then line l_3 is the required line.

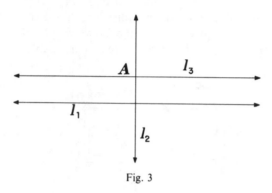

Fig. 3

In making constructions like those previously mentioned, there are two general rules. All lines drawn along an edge of the Mira should be drawn along the rebated edge, and only one piece of paper should be placed between the Mira and the flat, undersurface. If several pieces of paper are underneath the paper being used, the Mira tends to give inaccuracies, particularly when the construction requires that one end of the Mira be off the stack of papers.

Congruency of two coplanar circles can be investigated by attempting to make the image of one circle correspond to the other circle. In the case of other coplanar figures such as segments, triangles, and quadrilaterals, congruence can be determined when one figure is the image of the other under a line reflection.

A sequence of Mira activities like those described here have been constructed by the author and used in workshops for elementary teachers from the area surrounding Clarksville, Tennessee. The reaction to

these workshops has been very favorable. Many of the teachers have in turn used revised versions of the activities in their own classrooms. These activities have been received enthusiastically and teachers report that in addition to helping students learn geometric concepts, the activities are fun for the children. In classroom situations some teachers have used Miras in large group work while others use them for learning center work.

Miras are available commercially from Creative Publications and the Cuisenaire Company of America for about $3 each, but they can be made for approximately one-third of that, excluding labor. Two publications, *Mira Math for Elementary School* and *Mira Activities for Junior High School Geometry*, are also available from the same distributors. These workbooks contain directions on how to use the Mira in addition to many interesting and challenging Mira activities.

Geoboard activities for primary grades

RICHARD O. KRATZER and
BRUCE A. ALLEN

An associate professor in the College of Liberal Arts and an
assistant professor in the College of Education, respectively,
at the University of Maine at Portland—Gorham, Richard Kratzer and
Bruce Allen have worked closely together during the past three years
in developing a mathematics curriculum laboratory. They teach
mathematics methods courses in a public school setting where
the preservice teacher has an opportunity to interact with
pupils and experienced teachers.

The geoboard is a popular device for developing a variety of geometric and arithmetic patterns in mathematics. It frees the child from the static pictures and diagrams of the traditional paper-and-pencil activities. When children are able to "stretch" a rectangle, form squares of many different sizes, and transform one shape into another, they have a feeling of creating their own geometry. With an aroused interest children are more apt to sense patterns and shapes that occur in their own explorations.

Teaching with the geoboard is very much in keeping with the Piagetian notion that children should manipulate objects during the learning process. Through their manipulations of hands-on materials, the students will begin to perceive ideas that can be abstracted from the concrete devices. The discoveries that are made or at least sensed by the children because of their own involvement are more likely to be retained longer than if they were told about the concepts related to the discoveries.

One geoboard activity that is appropriate for primary-grade children involves a geoboard with a 5-by-5 array of nails, a supply of elastics (preferably colored), and a deck of square cards. The edge of each square card is slightly less than two geoboard units in length. Each card has a hole punched through its center so that when it is mounted on a geoboard nine nails can be easily identified, the nails around the perimeter of the square and the nail that protrudes through the hole in the center of the card. (Fig. 1) On one side of

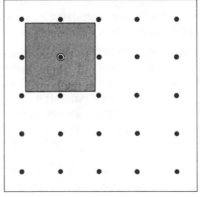

Fig. 1

each card, a simple straight-line design can be drawn using the positions of the nine nails as vertices of the design. (See figure 2 for several examples.)

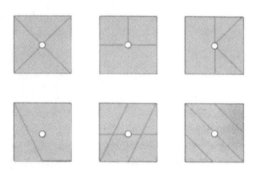

Fig. 2

Activities with the cards can be used in the primary grades in a sequence of developmental steps.

Step 1: Have the child mount a card in the upper left corner on the geoboard and then, with the elastics, reproduce the design pictured on the card. This step provides an opportunity for the child to develop fine motor skills as well as to become familiar with the ideas of points and lines.

Step 2: After the child has reproduced a design as described in step 1, have the child reproduce the same design to the right of the mounted card. (Often the child will reproduce a mirror image of the first design.)

Step 3: Lay a card to one side of the geoboard, and then have the child reproduce the design on the geoboard. This step promotes perceptual-motor match skills, which are important to writing numerals and letters.

Step 4: Show a card to the child, remove it from his sight, and then have him reproduce the design on the board. This activity aids in developing the children's ability to reconstruct a design from a mental image, which is a prerequisite to writing words and multidigit numerals.

In steps 3 and 4, it is helpful if a card with no design is mounted on the geoboard to serve as a "frame of reference" for the child when he is attempting to reproduce a design on the geoboard. Since the main objective of the activities is to provide readiness experiences for children, it is preferable for pupils to work with the cards at their leisure rather than to rush through the four recommended stages. The sequence pace should be adjusted to meet the needs of the class.

When children have gained some proficiency in reproducing simple straight-line designs, a set of cards that picture simple polygons can be used. (See figure 3.)

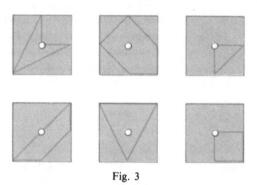

Fig. 3

A related activity that may be suitable for grades two and three would be to have the children make their own designs on the cards and then reproduce their designs on the geoboard; or to create designs on the geoboard and then reproduce the designs on the cards. The attribute of color, as well as shape, may also be developed by producing designs of various colors and having the children reproduce each design with elastics in the same color scheme as the design.

Another readiness activity with the geoboard that will help to develop reading and writing skills involves having children place small tags (with holes punched in their centers) on the geoboard according to certain directions. For example, the following directions might be given:

Place a tag on a nail in the bottom row.
Place a tag on a corner nail.
Place a tag on the middle nail.

These readiness activities can be further extended to develop and reinforce the children's vocabulary. For example, have the child stretch an elastic across the middle row of nails and then ask, "Can you place a tag above the elastic?" Then, without changing the elastic, remove the tag, turn the board one-quarter turn, and ask the child to place a tag on any nail that is on the left side (or the right side) of the elastic. Or the child can be directed to form a square with an elastic on the geoboard and then asked to place a tag inside the square, on the square, or outside the square. Similar questions could also be asked during the activities involving the cards mounted on the geoboard. The same activities can be made more sophisticated by using tags of several different colors. The children can be directed to place a red tag on a corner peg, or a blue tag on the second peg of the bottom row. Here the children are being asked to complete a task with two or three directions given in one set of instructions.

An activity that promotes thinking strategies is a variation of tic-tac-toe. Here two children can play a game where they take turns alternately placing tags on nails and trying to get four of their own tags in a row or column or diagonal. Will the first player always win? Is it better to start on the margin or on the center peg?

We have described several geoboard activities that provide opportunities for developing readiness for learning as well as for acquainting children with vocabulary and patterns important to geometry and arithmetic. It is hoped that the reader will be encouraged to use these geoboard activities with children in the primary grades.

Let's Do It

Let's Take a Geometry Walk

By **Glenn Nelson**
University of Northern Iowa
Cedar Falls, Iowa
and **Larry P. Leutzinger**
Area Education Agency 7
Cedar Falls, Iowa

Showing the relevance of a topic in mathematics to life can make the topic more meaningful to students. Noting the topic's existence and use in the world around us can provide motivation for the study of the topic and opportunities for real life applications. The study of geometry is a good example. Enterprising teachers have always used objects in the classroom to illustrate geometric terms and ideas. This approach can be extended by taking a geometry walk outside the classroom.

To promote a greater awareness of geometry in the real world, encourage students to note that (1) an object's function or use may determine its shape, (2) aesthetics or appearance may determine its shape, and (3) the vocabulary of geometry may be helpful in describing shapes and spatial relationships. A projector screen, for instance, is rectangular because the images displayed on it are generally rectangular and a rectangular screen is easier to roll up than a circular screen would be. An abstract painting or sculpture may use a shape, or combination of shapes, because a particular shape is pleasing in appearance. Many things, such as buildings or furniture, may be shaped as they are because of both function and aesthetics.

As you plan to explore your school building, schoolyard, and nearby neighborhood, make a note of questions that will direct students' attention to the large role that geometry plays in their lives. Questions should focus not only on shape recognition and use of geometric vocabulary but also on how function and aesthetics determine shape. Sketch a map for your planned walk and mark on it the places where, on your walk, you will ask certain questions. Plan your schedule so that some class time will be available after the walk for a discussion of the students' observations.

Following are samples of questions that you might ask students (and some possible responses). You will be able to think of other questions that are more appropriate for your circumstances and surroundings. Although divergent, meaningful responses to questions about function and aesthetics can be made by youngsters of widely differing ages and abilities. The "correctness" of learners' answers to questions on vocabulary may be dependent on the amount of geometry they have studied in school. You will need to plan your questions accordingly.

Functional Shape

1. *Why is each opening in the bicycle rack a long, narrow rectangle (fig. 1)?* "To keep the wheel and bicycle from falling over." "So the wheel will fit right in." "If it was a circle, you couldn't fit as many bicycles in the rack."

2. *Why is a bicycle wheel shaped like a circle instead of a square?* "So it's easier to pedal." "It makes the ride a lot smoother." *Pretend that you are riding a bicycle with square wheels. Show me how you would look.*

3. *Why is each end of the frame of the swing set shaped like a triangle (fig. 2)?* "Because it makes it steady, so you won't tip it over." *If we made the base of each end of the frame shorter, would it be as steady?* "No, when you swing way out you could tip it over."

4. *Why is a football not round like other balls?* "It would be harder to pass, harder to grip." "This way it's easier to cradle it in your arm and carry it without fumbling."

5. *Look at that factory's smokestack. Why is it cylindrical?* "Because you get a bigger smoke hole with fewer bricks."

Aesthetic Shape

1. *Notice the differences between the shapes found in that older building and the shapes found in that newer building. In which building would you prefer to work?* "The older one. I like all those perpendicular line segments; it looks more solid." *Which building would you prefer to draw?* "The newer one, 'cause all those triangles make it more interesting."

2. *Do you like the shape of that car parked over there?* "No, it's too boxy-looking." *How would you redesign its shape to look better?* "I'd make it lower in front and kind of slope it back more to make it look faster."

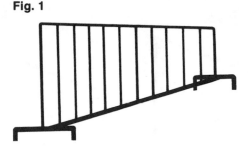

Fig. 1

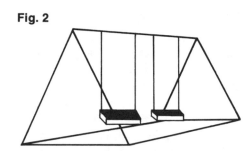

Fig. 2

3. *Many of the office signs we can see are rectangular, but this one is elliptical or oval-shaped. Why do you suppose Dr. Jones chose this shape for his sign?* "It looks nice and it's different. Maybe it's to take your mind off your pain."

4. *What do you see in that cloud formation? What does it look like to you (fig. 3)?* "The state of Texas." "My Aunt Barbara."

5. *Look at those two trees, the hemlock and the ash. Which shape do you like better (fig. 4)?* "The hemlock, 'cause its boughs are bent toward the ground." "I like the way the big branches of the ash reach toward the sky."

Geometric Vocabulary

1. *What kind of angle is formed here by this guy wire and this telephone pole?* "Acute."

2. *Which of the letters in our school's name, here on the wall, have line symmetry (fig. 5)?* "H, U, M, B, E, T, C, O." *Which have a horizontal line of symmetry?* "H, B, E, O." *Which have a vertical line of symmetry?* "H, U, M, T, O." *Which have both a vertical line of symmetry and a horizontal line of symmetry?* "H, O."

3. *Which of the letters in the school name would be examples of closed curves?* "O AND B."

4. *From here I see a large sign, shaped like a regular octagon. What*

Fig. 3

Fig. 4

Fig. 5

Fig. 6

color is it and where is it? "It's red and it's the stop sign at the corner."

5. *What shape is that yield sign (fig. 6)?* "A triangle."

Fig. 7

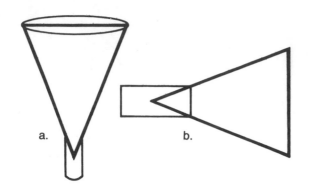
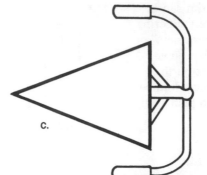
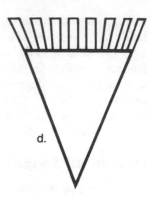

a.　　b.　　c.　　d.

After the Walk

Back in the classroom, list and discuss the instances of geometry that the class saw. Follow-up activities provide a challenge to all students and give your more creative thinkers an opportunity to extend themselves. The following are examples of follow-up activites that might be used with a class.

1. Have each student draw the outline of a shape that he or she observed on the walk. Share some of these with other members of the class. See if they can recall seeing this or a similar shape.

2. Have each student think of an object that is a certain shape because that shape helps the object do its job. A funnel would be a good example of this. Have the students sketch the shape of the object—maybe one or more of its faces or its cross-section. Each student can write what his or her object is and how its shape is useful in performing its job. Share the sketches with the class and ask for other objects for which a particular shape would be well-suited. For instance, an acute isosceles triangle could represent not only the cross-section of a funnel (fig. 7a) but also an ice scraper (fig. 7b), a bicycle seat (fig. 7c), or the head of a rake (fig. 7d).

3. Your students might also speculate about what shape or shapes an object might be if it were to perform a stated function. For example, you might ask students to sketch the design of a new and different fingernail file, doorknob, desktop or window, and write why they chose the shape they did. It might be interesting to see if they come up with some function-shape relationships that have already been found quite useful—the triangle for rigidity and stability, the circle for rolling and rotating, or the rectangle for tessellating or fitting together to cover a surface (fig. 8).

4. You might select a particular shape and ask students to use that shape, predominantly, to draw something that looks pleasing to them. You could be more specific and ask them to sketch a design for a piece of sculpture, a piece of jewelry, a sign or, combining function and aesthetics, an attractive solar panel, piece of furniture, or eating utensil. Allow them to choose the shapes they wish to use. Luckily, not everyone likes what someone else likes. Individual tastes may provide a variety of original creations that can be shared with the whole class. Ask the students which shapes they like best. Shapes may be associated with a product or event. A cone (or triangle), for instance, may remind one of an ice cream cone or a Christmas tree. Others may find just the shape alone appealing. You might show a golden rectangle (fig. 9), whose length is approximately 1.6 times its width, and state that many people find this the most pleasingly-shaped rectangle. Or ask students to draw the rectangle that is most pleasing to them.

Fig. 9

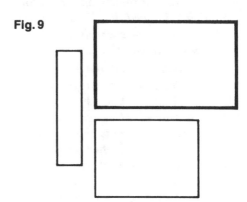

5. Make a display of three or four bottles of different shapes and ask students which they find the most attractive. If the bottles all have the same capacity, you might ask if any were designed so as to look as if they held more than others.

6. A geometric scavenger hunt could provide an opportunity to find objects that illustrate geometric terms. Groups of students could be given a list of terms and told to find and record the names of objects in the classroom that are examples of those terms. A rectangle, for example, could be found on the top of a table, a window, or a book. A circle could be seen in the face of a clock.

A geometry walk and the related activities should provide for your class some answers to the question, Why do we have to study geometry? And your students will begin to see that geometry *can* be relevant to everyday activities and interesting. □

Fig. 8

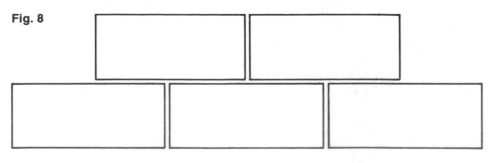

Some geometry experiences for elementary school children

J. PAUL MOULTON

An assistant professor of curriculum and instruction at Temple University in Philadelphia, Paul Moulton formerly taught at the University of Chicago Laboratory School, and is one of the authors of a new series of elementary school texts.

Geometry in the elementary school has been plagued with three pervasive emphases: (1) false utilitarianism, (2) superficial attention to form, and (3) sterile Euclidean formalism. For example, a great deal of attention is paid to the computation of areas, perimeters, and volumes with the idea that students need to be able to make these computations in order to buy land, canned goods, fencing, rugs, and other items sold by area, volume, or length. A large amount of time is also devoted to categorizing objects into classes such as isosceles, scalene, convex, closed, acute, and the like without any time having been spent on a study of the broader significance of the categorizing properties. And a lot of effort is put into teaching various ruler-and-compass constructions at a time when the philosophical justification for these impractical constructions is unknown to the students. The student whose geometry instruction has been pervaded with these emphases finds the subject dull and not particularly useful. This is unfortunate because the natural world and the man-made world alike abound with interesting objects in which geometric properties play a fundamental role in effecting the proper function of the object. The student ought to discover this.

Take, for example, the shafts by means of which Philadelphia fire hydrants are turned on and off. These shafts, when viewed from the top, are Reuleaux triangles, which are made up of three arcs with common radii and with centers at the three vertices. (See fig. 1.) A worth-

Fig. 1

while activity for an elementary student would be for him to actually go out into the neighborhood to observe that the shafts in the fire hydrants are shaped in this way, and then, back in his classroom, to figure out why that shape has been used. It will turn out that the shaft has been designed to keep children and other non-firefighting types from turning the hydrants on. A Reuleaux triangle is a curve of constant width, so that an ordinary wrench with parallel jaws will simply slip around the triangle as it would around a circle. (See fig. 2.) It takes a special wrench—supposedly available only to firemen—to turn the shaft of the hydrant.

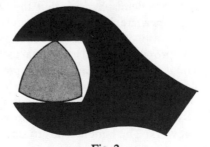

Fig. 2

Fire hydrants are not the only place where curves of constant width have a useful application. Gardner (1963) describes several. Children might be asked to suggest other uses, and certainly they could make cardboard models with which they can experiment. Children are not likely to produce any marketable patents in the process, but they are very likely to come away with a better appreciation of the role of abstract geometric properties in determining the design of practical objects.

Another experience that is easy to provide is to examine the design of the common hemispherical mixing bowl, and to determine why this shape has been chosen. Two features, which students can readily identify for themselves, make a bowl of this design more practical than some other. One is that a hemisphere has constant curvature; a bowl made in this shape has no sharp grooves or corners. Because there are no hard-to-reach places, food does not get trapped in corners and left out of the mixing process, and when the time comes to clean the bowl, every part of its surface is equally accessible. The other feature is that a hemisphere holds more for a given amount of surface than does any other open shape. Choosing this shape provides a bowl that is as light as possible and that has the least surface to which the mixture can cling—both very practical considerations for a cook.

Other instances in which the constant-curvature property or the minimal-surface property of the hemisphere plays a role can be found, and children can be asked to look for them. For example, children may observe that bubbles on the surface of a liquid are hemispherical, that the two halves of a pea are hemispherical, and so on. In the case of the pea, the constant-curvature property permits the pea to be rolled about by the wind and thus to spread the colony of pea plants from which it came. In the case of the bubble, it is the minimal-surface property that is most obviously at work. The bubble tends to shrink, but because of the air trapped inside, it can shrink only so far. The shape that allows it to shrink the most is that of the hemisphere.

A third example of a geometric property that is extensively employed in the man-made world is the side-side-side property of congruent triangles; that is, the property that asserts that the shape of a triangle cannot change unless the length of at least one of its sides changes. Perhaps the most impressive application of this property is to be found in the long booms of the building cranes used to lift heavy loads. Although these booms are made of relatively slender pieces of steel, they are able to retain their shape even when supporting huge buckets of concrete, massive pieces of machinery, or other heavy loads. If construction happens to be going on near the school, children can be taken to the site and asked to sketch what they see. Back in their classroom they can then attempt to build similar booms out of paper drinking straws. One-inch lengths of pipe cleaner stuck into the straw ends with a drop of cement make excellent joints. (See fig. 3.) Properly designed

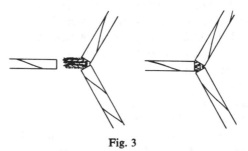

Fig. 3

models will prove to be amazingly rigid and strong, whereas models that fail to exploit the rigidity of triangles will not. A lesson on triangles taught in this way will be far more fun, meaningful, and impressive than, say, a lesson in which students work only with drawings. This lesson, too, can be followed by others in which children find instances where triangles have been employed in the design of common objects in order to ensure their rigidity. Examples will be found in bridges, shelf brackets, garage doors, telephone-pole crossarms,

Fig. 5

and elsewhere.

Another extremely useful and commonly exploited geometric property is the one that relates the circumference of a circle to its radius: as the radius increases, the circumference increases, and conversely. This property accounts for the strength of stone arches, for the stiffness of venetian blind slats, for the locking and unlocking of slide-fastener teeth, and for the satisfactory performance of countless other natural and man-made objects.

The stone arch doesn't collapse because to do so it would have to shorten its radius of curvature, and if it were to do this, it would have to shorten its circumference. Since the stones are incompressible, the circumference cannot grow shorter, the radius cannot grow shorter, and the arch stays up. (See fig. 4.) Children can discover

Fig. 4

this by building arches with blocks that they cut from styrofoam or similar material. Building arches with styrofoam blocks is a very impressive experience, and one that considerably enhances a child's appreciation of such marvels as the Roman aqueducts and medieval cathedrals, which were built without steel or other brute-strength materials.

However, it is not necessary to study something as elaborate as a self-supporting arch to see this principle. It can be demonstrated with an ordinary strip of paper. Held by one end, the paper strip will flop; held by both ends, it will sag. If the strip is folded along its entire length, however, the strip becomes rigid like the slats in a venetian blind. (See fig. 5.) It cannot sag because if it did the radius to the fold line would be shorter than the radius to the outer edges, and the length measured along the fold would have to be

shorter than the length measured along the edges. The paper would have to stretch along the edges. Since the paper does not stretch appreciably, the folded strip does not bend.

The foregoing are only a few of the many examples in which the ingenious employment of geometric principles results in a well-designed product—where mathematics consists of something beyond a dull, sterile assimilation of facts and skills. We turn out a substantial proportion of our students with an intense dislike of mathematics, and an even larger proportion with a belief that mathematics consists mainly of computation and of meaningless preoccupation with impractical concerns, such as ruler-and-straightedge constructions, commutative laws, and proofs. In every home and in every neighborhood, there is a generous supply of common objects which students can observe and in which mathematical properties are to be found. If students can have the kinds of experiences described here and more, then we won't have such a high proportion of mathematically naive and emotionally turned-off students leaving our classrooms.

Reference

Gardner, Martin. "Mathematical Games: Curves of Constant Width." *Scientific American* (February 1963): 148–56.

Let's Do It

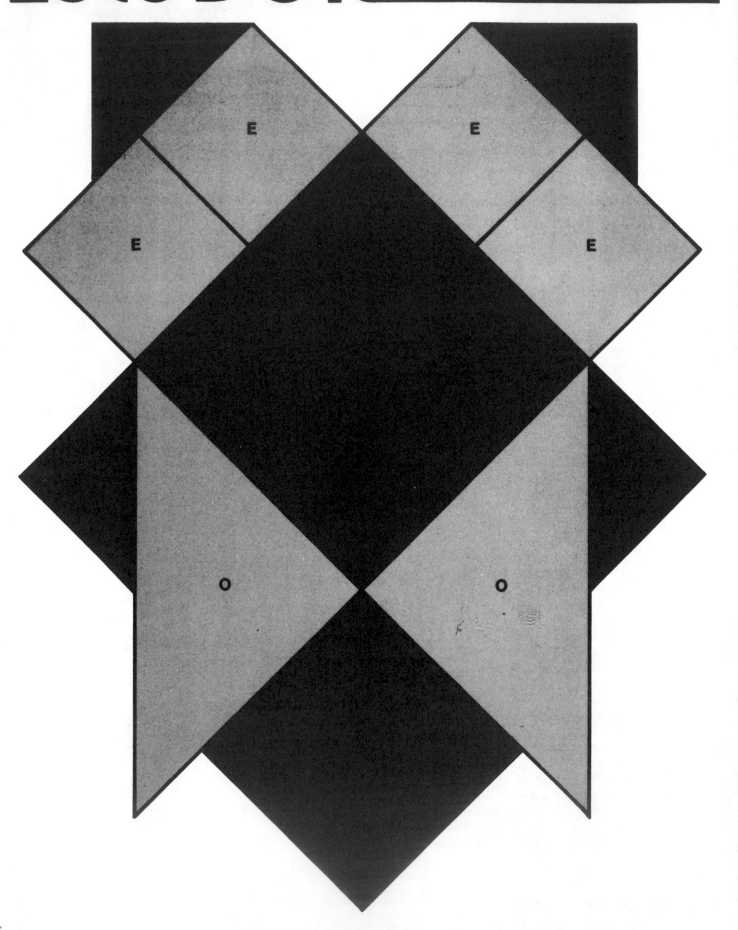

Many of the best problem-solving situations in primary schools come from everyday situations: "How many more chairs will we need if we are having five visitors and two children are absent?" "How many cookies will we need if everyone has two?" However, these situations do not always arise at the appropriate time, nor do we always have time to take advantage of them when such situations do arise. There is a need for a set of problems that can be used at any time and the problems selected for this article are of this type. They require a minimum amount of computation but, often, a maximum amount of thinking.

All the problems use a set of geometric pieces. The problems involve not only geometry, including area, but also logic, combinations, division, and money. None of these topics is dealt with formally; all are presented in the form of a puzzle or problem.

The pattern for the five easy pieces is shown in figure 1. There are four *A*s, four *E*s, one *I*, two *O*s, and one *U*, or twelve pieces altogether. The problems are written for yellow and blue pieces. If you are making classroom sets, make a duplicating master and run off copies on yellow and blue construction paper. If you want more durable sets, use heavy tagboard. You might also want to make and use only one complete yellow and one complete blue set with a class.

Fig. 1

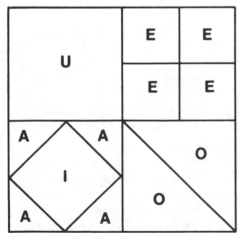

The problems can be put on four-by-six-inch index cards from which you can make up sets suitable for your children. Or you may wish to use the problems in other ways—oral directions to

Problem Solving With Five Easy Pieces

By **Mary Montgomery Lindquist**
National College of Education
Evanston, Illinois

the children or a problem each week posted on the board. The pieces that are needed for each problem are indicated; "one complete set" refers to either one yellow or one blue set. The answers to the problems are included.

As you and the children work with these activities, you will see other problems or questions. For younger children, you may want to make more of the problems they find within reach (probably the simpler puzzles, patterns, and cost cards) and for older children you will want to extend the problems.

Basic Relationships

In these three sets of problems, the children are asked to investigate the area relationships among the pieces. Doing these problems, especially those in Basics 1, should help the children do the later logic and cost problems.

In Basics 1, most children will have no difficulty seeing that two *A*s cover an *E*; more children will have trouble seeing that four *A*s cover an *I* as well as an *O*. There are not enough *A*s to completely make a *U*, but with a little ingenuity, children should be able to see that it takes eight *A*s. Notice how the children solve the problems. Do any cover as much as they can with *A*s and see that half of *U* is covered? Do any of the children use *E*s to help to solve the problem? Do any use *O* to help?

Basics 1
Take: One complete set
How many *A*s does it take to cover

an *E*?
 I?
 O?
 U?
Answers: *E*, 2; *I*, 4; *O*, 4; *U*, 8

Since the *E*s will not completely cover *I* or *O*, the problems in Basics 2 are more difficult than those in Basics 1. The children will either have to imagine cutting up piece *E* or use the relationship they found in Basics 1—since two *A*s cover an *E*, and four *A*s cover an *I*, it would take two *E*s (if they were cut) to cover *I*.

Basics 2
Take: One complete set
How many *E*s does it take to cover an *I*?
 O?
 U?
Hint: You may have to pretend to cut piece *E*. Use *A* to help you.
Answers: *I*, 2; *O*, 2; *U*, 4

The problems in Basics 3 reinforce the relationship that *I* and *O* are the same in area, and *U* is twice as large as *I* and twice as large as *O*.

Basics 3
Take: One complete set
1. If you could cut up piece *I*, would it cover piece *O*?
2. How many *O*s does it take to cover *U*?
3. How many *I*s (pretend that you could cut an *I*) does it take to cover *U*?
Answers: 1, yes; 2, 2; 3, 2

Puzzles

Do not expect all children to come up with all the variations on Puzzles 1. As the children find them, you may want to put the possible arrangements on the bulletin board.

Puzzles 1
Take: One complete set
How many different ways can you make a copy of the square *U*?
 Example:

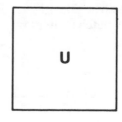 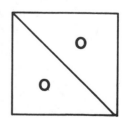

Answers:

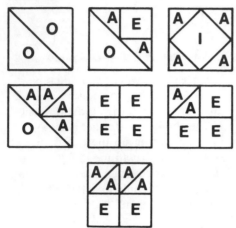

Other arrangements of the pieces are possible in some of these squares.

Only one solution for Puzzles 2 is given here; other solutions are possible. You may vary this puzzle by asking the children to make one triangle, one rectangle, or one parallelogram.

Puzzles 2

Take: One complete set
Take all the pieces. Can you make one large square?
Answer:

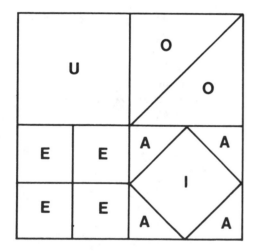

One solution. Others are possible.

Let children experiment with Puzzles 3 through 6 and make up other puzzles. For young children, put only one question on a card. Be sure children understand that in all these puzzles, when two pieces are fit together, sides of equal length must be matched. For example,

Puzzles 3

Take: Two *A*s
1. Can you make a triangle?
2. Can you make a square?
3. Can you make a four-sided figure that is not a square?
Answers:

1.   2.

3.

Puzzles 4

Take: Three *A*s
1. Can you make a four-sided figure?
2. Can you make a five-sided figure?
Answers:

1.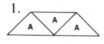

2.

Puzzles 5

Take: Four *A*s
1. Can you make a triangle?
2. Can you make a square?
3. Can you make a rectangle that is not a square?
4. Can you make a four-sided figure that is not a rectangle?
5. Can you make a five-sided figure?
6. Can you make a six-sided figure?
Answers:

1.   2.

3.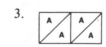

4.

5.   6.

Puzzles 6

Take: Four *E*s
1. Can you make a four-sided figure?

2. Can you make a five-sided figure?
3. Can you make a six-sided figure?
4. Can you make an eight-sided figure?
Answers:

1.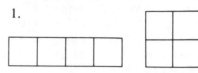

2. not possible

3.

4.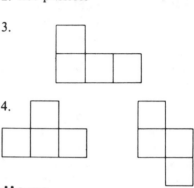

Patterns

In the pattern-making problems, children may come up with answers different from those given. Ask children to explain their patterns—their patterns, though different, may be legitimate. Also ask the children who come up with the given solutions to explain the patterns—they may be seeing a pattern in a different way.

Patterns 1

Take: Yellow and blue *A*s and *E*s
Put in a line:
1. yellow *A*, yellow *A*, blue *A*, yellow *E*.
What would be the next two pieces?
2. yellow *A*, blue *E*, blue *A*, blue *E*.
What would be the next two pieces?
3. yellow *A*, yellow *A*, yellow *E*, blue 'A*.
What would be the next two pieces?
Possible solutions:
1. yellow *E*, blue *E*; 2. *A*, *E* (either color); 3. blue *A*, blue *E*

As with the puzzles, the rule of putting pieces together by matching sides of equal length must be followed. The restriction that one side of one piece must be placed on a given line limits the endless possibilities of "tipping." The further restriction that the pattern must be above the line or, if problems are written on the cards, "on the card" is optional. Either eliminates the possibility of flipping the pieces over the line—getting a mirror image, in other words. Feel free to omit these restrictions, but if you do, you can expect many other possible results.

Patterns 2
Take: Two blue *A*s and two yellow *A*s
Using two pieces, how many different figures can you make? (One triangle must be on the line and the figure should be above the line.)
Example:

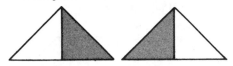

Answer:
Each shape can be varied four ways with the color changes.

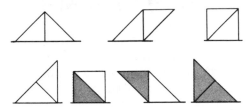

Patterns 3
Take: Three yellow *E*s and three blue *E*s
Using three pieces, how many different patterns can you make? (One square must be on the line and the pattern should be above the line.)
Example:

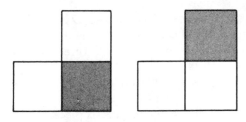

There are eight color variations of this shape.

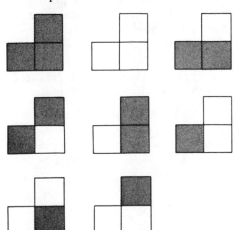

There are five other shapes. Each can be varied eight ways with color.

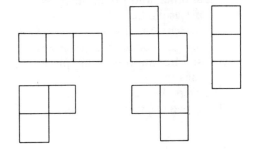

Patterns 4
Take: Four yellow *A*s and four blue *A*s
Take any four pieces and make a shape. How many variations of your shape can you make by changing colors?
Answer: Suppose the shape is

There are sixteen color variations

Some children may realize that once they specify a shape on Patterns 2, there are four ways to vary the color. Patterns 4 is a variation of the ideas begun on Patterns 2 and 3. If there are children who see that there are sixteen ways to vary the colors, you may want to look at Patterns 2, 3, and 4 with them. There are four ways to vary the colors with two pieces, eight ways with three pieces, sixteen ways with four pieces. What would you expect with five pieces?

Do not expect many children at this level to see the numerical pattern. You can be satisfied if they begin by moving the pieces to see the variety of possibilities. You can help those children who are ready to organize their work by asking questions like the following: How many variations can you make if no triangles are blue? If only one triangle is blue? If two triangles are blue? If three triangles are blue? If four triangles are blue?

Logic

The logic cards progress in difficulty from 1 to 6. It may help some children to write the new names for the pieces on slips of paper so they can move the names around to match the clues.

Logic 1
Take: One *E*, one *I*, one *U*
These three squares renamed themselves—Ali, Bet, and Tim.
Ali said, "I'm bigger than Tim."
Bet said, "I'm bigger than Ali."
Who's who?
Answers: Tim is *E*, Ali is *I*, Bet is *U*.

Logic 2
Take: One *A*, one *E*, one *O*
These three pieces renamed themselves—Bill, Jill, and Lil.
Bill said, "I'm twice as large as Jill."
Jill said, "I'm the same shape as Lil."
Who's who?
Answers: Jill is *A*, Bill is *E*, Lil is *O*.

Logic 3
Take: One *A*, one *E*, one *O*, one *U*
These four pieces renamed themselves—Mary, Larry, Harry, and Cary.
Mary said, "I'm a fourth as large as Harry."
Larry said, "I'm bigger than Harry."
Who's who?
Answers: Mary is *A*, Cary is *E*, Harry is *O*, Larry is *U*.

Logic 4
Take: One *A*, one *I*, one *O*, one *U*
These four pieces renamed themselves—Floe, Joe, Moe, and Woe.
Joe said, "I'm twice as large as Moe."
Floe said, "I'm the same shape as Woe."
Woe said, "I'm smaller than Moe."
Who's who?
Answers: Woe is *A*, Moe is *I*, Floe is *O*, and Joe is *U*.

Logic 5
Take: One *A*, one *E*, one *I*, one *O*, one *U*
These five pieces renamed themselves—Dan, Nan, Stan, Ann, and Fran.
Fran said, "I'm the same size as Dan, but larger than Stan."
Dan said, "I'm larger than Nan, but smaller than Ann."
Nan said, "I'm smaller than Fran,

but larger than Stan."

Fran said, "I'm the same shape as Nan."

Who's who?

Answers: *A* is Stan, *E* is Nan, *I* is Fran, *O* is Dan, *U* is Ann.

Logic 6

Take: One *A*, one *E*, one *I*, one *O*, and one *U*

These five pieces renamed themselves—Al, Cal, Mal, Pal, and Sal.

Pal said, "I'm twice as large as Sal."

Mal said, "I'm four times as large as Al and the same shape."

Pal said, "I'm larger than Mal."

Who's who?

Answers: *A* is Al, *E* is Cal, *I* is Sal, *O* is Mal, *U* is Pal.

Areas

Instead of having the children cover the figures with other pieces to find the area of the figures, they are given the results of someone's covering. The children then have to determine what units were used. If the children enjoy this challenge, make up some larger shapes that permit more possibilities.

Areas 1

Take: One complete set

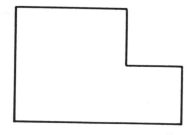

1. Bob covered this figure with pieces. He took 4 of one kind and 3 of another. What pieces did he use?
2. Karen covered this with pieces. She took 2 of one kind and 2 of another. What pieces did she use?

Answers: 1. four *A*s and three *E*s
2. two *O*s and two *A*s

Areas 2

Take: One complete set

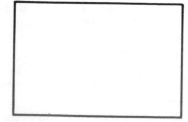

1. Hal covered this rectangle with pieces. He took 4 of one kind, 1 of another, and 1 of another. What pieces did he use?
2. Sue covered this with pieces. She took 2 of one kind, 2 of another, and 1 of another. What pieces did she use?

Answers: 1. one *O*, one *I*, four *A*s
2. two *O*s, two *A*s, one *E*

Costs

Costs 1 depends on the children seeing the basic area relationships between the pieces. It may help younger children to work with pennies (or chips).

Costs 1

Take: One complete set

1. If *U* costs 12¢, how much does
   E cost?
   A cost?
2. If *I* costs 8¢, how much does
   E cost?
   U cost?

Answers: (1) 3¢, 1½¢, (2) 4¢, 16¢

It may help in question 2 of Costs 2 to have the children paper clip the price on each of the four pieces.

Costs 2

Take: One *A*, one *E*, one *O*, and one *U*

1. If *A* cost 1¢, how much would each of the other pieces cost?
2. Using just these 4 pieces, how would you pay a bill for each of the following:
   1¢, 2¢, 3¢, 4¢,
   5¢, 6¢, 7¢, 8¢,
   9¢, 10¢, 11¢, 12¢,
   13¢, 14¢, 15¢

Answers: 1. *E*, 2¢; *O*, 4¢; *U*, 8¢
2. 1¢, *A*; 2¢, *E*; 3¢, *A* and *E*; 4¢, *O*; 5¢, *O* and *A*; 6¢, *O* and *E*; 7¢, *O*, *A*, and *E*; 8¢, *U*; 9¢, *U* and *A*; 10¢, *U* and *E*; 11¢, *U*, *E*, and *A*; 12¢, *U* and *O*; 13¢, *U*, *O*, and *A*; 14¢, *U*, *O*, and *E*; 15¢, *U*, *O*, *E*, and *A*.

Costs 3 is similar to the puzzles and the rule of fitting sides together applies. Again, for younger children you may want to separate the questions, one question to a card. For children who find these problems easy, change the price of *A* to 3¢, or make up cost varia-

tions for Puzzles 3 through 7.

Costs 3

Take: Four *A*s. Each *A* costs 1¢.

1. Can you make a four-sided figure that costs 3¢?
2. Can you make a three-sided figure that costs 4¢?
3. Can you make a five-sided figure that costs 3¢?
4. Can you make a five-sided figure that costs 4¢?

Answers:

1.   2.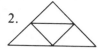

3.   4.

On Costs 4, the children have to focus on both the total number of pieces and the total cost.

Costs 4

Take: One complete set

1. If *A* costs 1¢, what would each of the other pieces cost?
2. Can you sell the pieces to two people so that each person gets the same number of pieces and each person would have to pay the same amount?

Answers:

1. *E* costs 2¢, *I* costs 4¢, *O* costs 4¢, and *U* costs 8¢
2. First person: one *U*, three *E*s, two *A*s
   Second person: two *O*s, one *I*, one *E*, two *A*s
   (Each person gets six pieces that cost 16¢.)

The ideas for many of the problems that have been described here come from *Developing Mathematical Processes* (DMP). Acknowledgement is due the many writers of DMP who inspired these problems and the many teachers and children who have tried similar problems. □

Polygons, Stars, Circles, and Logo

By **Bill Craig**

Fig. 1

REPEAT 8[FD 50 RT 45]　　　REPEAT 6[FD 50 RT 60]

Many teachers are excited about the potential uses of Logo with elementary school students. The language gives students access to mathematical topics they have not previously explored. The following activities use Logo in the study of geometry, number theory, and problem solving. The activities assume that students are familiar with turtle-graphics commands (FORWARD, BACK, RIGHT, LEFT) and know how to define procedures. The activities are designed for students in the upper elementary and middle school grades. The star procedure and explorations are adapted from *Discovering Apple Logo* by David Thornburg. The book contains excellent ideas for the use of Logo as a tool for mathematical explorations. See the Bibliography for additional resources.

The Rule of 360

Students are often introduced to the rule of 360 by using the REPEAT statement to draw regular polygons. The rule states that the turtle will turn a total of 360 degrees when drawing a regular polygon. A regular polygon is one that has all equal angles and all equal sides. After some introductory work with the REPEAT statement, students can use worksheet 1 to record the number and sizes of turns required to draw each of the regular

polygons. Figure 1 shows an octagon and a hexagon as drawn by the computer. Once commands have been established for several regular polygons, have students examine the results for patterns. The bottom of worksheet 1 focuses students' attention on the all-important rule of 360. Students need to understand that the input for the REPEAT when multiplied by the input for the turn equals 360. For example, the command for a square is

REPEAT 4[FD 50 RT 90]

and 4 · 90 = 360.

Make sure students can verbalize the relationship between the number of sides and the angle turned. The bottom of worksheet 1 contains further explorations with the rule of 360. Have students explain how they determined the size of the turn needed to draw a ten-sided figure (36 degrees) or a twelve-sided figure (30 degrees). See if anyone discovers that one way of

drawing a Logo circle is by drawing a regular polygon with many sides. Three possibilities include these:

REPEAT 360[FD 1 RT 1]
REPEAT 36[FD 5 RT 10]
REPEAT 20[FD 5 RT 18]

Challenge students to see how many circles of different sizes they can draw.

This investigation can be extended by introducing variables through the use of a polygon procedure:

TO POLY :SIDES
REPEAT :SIDES [FD 40
　　RT 360/:SIDES]

POLY 9 means this:

REPEAT 9 [FD 40 RT 360/9]

POLY 5 creates a shape with the following directions:

REPEAT 5 [FD 40 RT 360/5]

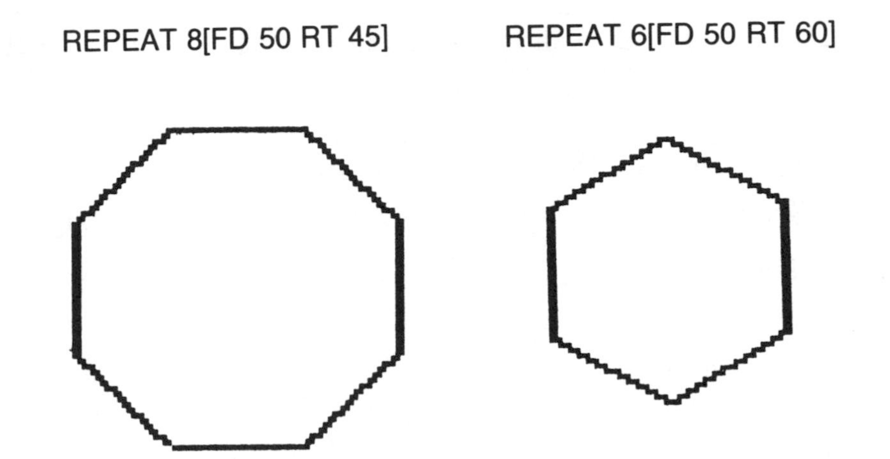

Bill Craig is currently a computer-education program specialist for Chesterfield County Schools, Chesterfield, VA 23832. He has an interest in the use of Logo to enhance mathematics instruction in the elementary schools.

THE RULE OF 360

WORKSHEET 1

1. Fill in the blanks to draw each of these regular polygons:

Triangle	REPEAT	________	[FD 50	RT	________]
Square	REPEAT	________	[FD 50	RT	________]
Pentagon	REPEAT	________	[FD 50	RT	________]
Hexagon	REPEAT	________	[FD 50	RT	________]
Octagon	REPEAT	________	[FD 50	RT	________]

2. Look over your results. What patterns do you see in the blanks you filled in?

3. Record the following information for each of the regular polygons:

Shape	Number of Sides	Angle Turned	Sides × Angle
Triangle			
Square			
Pentagon			
Hexagon			
Octagon			

4. What additional patterns do you see now?

5. How would you draw a ten-sided regular figure?

6. How would you draw a twelve-sided regular figure?

7. How would you draw a circle?

From the *Arithmetic Teacher*, May 1986

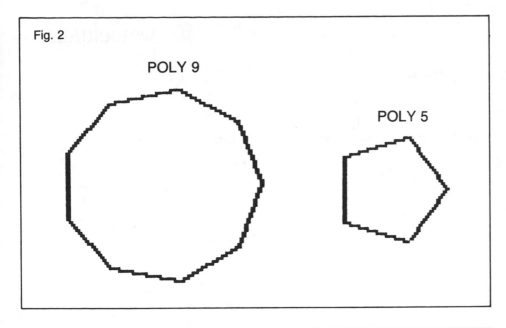

Fig. 2

POLY 9

POLY 5

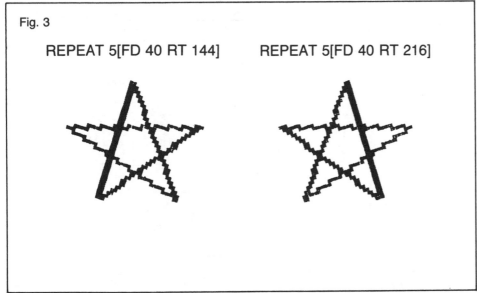

Fig. 3

REPEAT 5[FD 40 RT 144]

REPEAT 5[FD 40 RT 216]

These shapes are shown in figure 2. Have students try this procedure with many different inputs. Make sure that they understand how and why the procedure works.

Most teachers who use Logo engage students in similar activities. But the mathematical value of these exercises should not be underestimated. Students completing worksheet 1 are—

1. discovering properties of regular polygons;

2. finding patterns;

3. developing a generalization (rule of 360); and

4. applying the generalization to a new problem (drawing a circle).

Logo and the Stars

Now go back to the command for the pentagon. Double the turns so that it is now

REPEAT 5[FD 40 RT 144].

Direct students' attention to the new shape, that is, a five-pointed star. Now triple the original turn, REPEAT 5[FD 40 RT 216]. As shown in figure 3, this command also produces a five-pointed star. Challenge students to determine whether doubling and tripling of the turn of a regular polygon will always produce a star with the same number of points as the original polygon.

Try a hexagon: REPEAT 6[FD 40 RT 120]. The result is merely a triangle drawn twice. How about tripling the turn so that it is now REPEAT 6[FD 40 RT 180]? Ask students if they can explain why this command yields a series of straight lines because of the turn of 180 degrees.

Some of the changes produce stars, whereas others do not. A variable procedure will allow students to check many possibilities quickly.

```
TO STAR :SIDES :MULT
REPEAT :SIDES[ FD 40
       RT (:MULT * 360/:SIDES) ]
END
```

The first input (:MULT) tells us how many times to multiply the angle; the second (:SIDES) determines the number of sides.

With the use of worksheet 2, have students run the procedure with many different inputs and record their findings. The chart on worksheet 2 will help lead students to the discovery that pairs of prime or relatively prime inputs create stars. Direct students' attention to the results for figures with sides of 3, 5, 7, 11, or 13, which are prime numbers. Except for the multiplier of 1, these shapes are always stars. Examine the multipliers for sides of 6, 8, or 9 that create stars. Stars are created here when the multiplier has no common factors with the number of sides.

The more one experiments with this procedure, the more patterns and avenues for investigations are discovered. Students can also search for patterns in entries that create straight lines, triangles, or squares.

Once again, students are hypothesizing, testing, and predicting, as well as manipulating variables and exploring number theory in a different setting.

Logo and Symmetry

This activity uses the same STAR procedure. To see better some of the figures created in this figure, you might want to increase the input for FORWARD:

```
TO STAR :SIDES :MULT
REPEAT :SIDES [FD 50
       RT (:MULT/:SIDES)]
END
```

```
TO STAR :SIDES :MULT
REPEAT :SIDES[RT (:MULT * 360/:SIDES)]
END
```

1. Try the STAR procedure with each of these inputs.
   Record whether or not a star is drawn. Add some of your own inputs.

:SIDES	:MULT	Star? (yes/no)
5	2	_____
5	3	_____
6	2	_____
6	3	_____
6	4	_____
7	2	_____
7	3	_____
7	4	_____
7	5	_____
8	2	_____
8	3	_____
8	4	_____
8	5	_____
8	6	_____
9	2	_____
9	3	_____
9	4	_____
9	5	_____
9	6	_____
11	2	_____
11	3	_____
11	4	_____
11	5	_____

2. Which pairs of inputs give stars?

3. Which pairs of inputs do not give stars?

4. Which inputs for :SIDES always create stars? What kinds of numbers are these?

5. Predict what the following inputs will create:

   Prediction

   STAR 2 9 _____   STAR 4 9 _____
   STAR 3 9 _____   STAR 5 9 _____
                               STAR 6 9 _____

   Check your predictions by running the procedures.

6. Can you write a rule that will help you make directions for creating stars?

From the *Arithmetic Teacher*, May 1986

```
TO  STAR  :MULT  :SIDES
REPEAT  :SIDES [FD 50 RT(:MULT * 360/:SIDES)]
END
```

1. Type:

STAR 1 5
STAR 4 5

   You have created a shape with line symmetry. Identify one or more lines of symmetry.

2. Type this pair of procedures to see another design with line symmetry:

STAR 3 8
STAR 5 8

3. Try some pairs of procedures on your own and record "yes" for those pairs that create a symmetrical design and "no" for those that do not.

Procedure	Yes	No
STAR _____ _____	_____	_____
STAR _____ _____		
STAR _____ _____	_____	_____
STAR _____ _____		
STAR _____ _____	_____	_____
STAR _____ _____		
STAR _____ _____	_____	_____
STAR _____ _____		
STAR _____ _____	_____	_____
STAR _____ _____		

4. What do you notice about the inputs that create symmetrical designs?

5. Which of these pairs will create symmetrical designs?

   a) STAR 4 9 _____   c) STAR 3 4 _____
      STAR 1 9   STAR 1 4 _____

   b) STAR 3 6 _____   d) STAR 5 11 _____
      STAR 5 6   STAR 6 11 _____

6. Name two other STAR procedures that will create a symmetrical design:

STAR _____   STAR _____

7. Use STAR to create a design with only (a) one line of symmetry, (b) two lines of symmetry, and (c) infinitely many lines of symmetry.

8. Try combining several different STARS with a turn between them, for instance:

STAR 2 3
STAR 1 3
RT 180
STAR 2 5
STAR 3 5

   Write the commands for your favorite design.

From the *Arithmetic Teacher*, May 1986

GEOMETRY IN UNUSUAL WAYS   161

Type the following: STAR 1 5. Then, without clearing the screen, type STAR 4 5. Ask students to describe what they see on the screen. Explain that they are looking at a shape with line symmetry. If a symmetrical figure is folded along a line of symmetry, the halves match exactly. Identify a line of symmetry on the computer's screen. Test the following pairs of inputs for line symmetry: STAR 3 8, STAR 5 8 and STAR 3 7, STAR 4 7. Have students identify lines of symmetry in the new shapes.

Not all pairs of STARS create a symmetric design. For instance, STAR 4 7 and STAR 5 7 create a design without any lines of symmetry. See figure 4 for examples of STAR designs. Have students use worksheet 3 to try many different pairs of STARS and record whether a symmetric design is created. Once the results have been examined, students should see that pairs of procedures in which the sum of the two inputs for :MULT equal the :SIDES input result in a symmetric design. For instance, STAR 4 9 and STAR 5 9 create a symmetric design (4 + 5 = 9), whereas STAR 3 9 and STAR 4 9 do not. The last questions on worksheet 3 provide an opportunity for a more open-ended exploration. Some very interesting designs can be created and examined. Students can benefit from producing printed copies of designs created here.

Summary

The emphasis of these activities is not on the learning of Logo (or programming) but on the learning of mathematics. The computer is used as an electronic manipulative that gives students another medium with which to explore and discover mathematics. The challenge for all teachers using computers is to develop a curriculum in which the computer facilitates the exploration of a discipline, either mathematics or another.

Bibliography

The following resources focus on the use of Logo as a tool with which to learn. Although all the books introduce the reader to Logo, they go beyond introductory work with the turtle to illustrate applications of the language.

Bearden, Donna, Kathleen Martin, and James Muller. *The Turtle's Sourcebook*. Reston, Va.: Reston Publishing Co., 1983.

Thornburg, David P. *Discovering Apple Logo*. Reading, Mass.: Addison-Wesley Publishing Co., 1983.

Tipps, Steve. *Nudges: IBM Logo Projects*. New York: Holt, Rinehart & Winston, 1984.

Watt, Daniel. *Learning with Logo*. New York: McGraw-Hill Byte Books, 1983. ♥

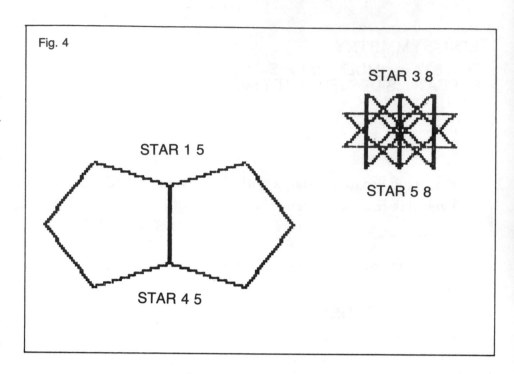

Fig. 4

STAR 3 8

STAR 1 5

STAR 5 8

STAR 4 5

Spatial Abilities

WHAT people perceive—how they see the world they live in—is influenced by experience. One of the purposes of formal education is to provide experiences for children that will help them reach their full potential as adults.

Progress in science and technology has made it possible to "see inside the brain." More is now known about what goes on inside the brain as children and adults learn and think. There is evidence to suggest that mathematical ability is correlated with spatial visualization, a function of the right side of the human brain. The two articles in this section relate activities to the development of spatial abilities. In "Improving Spatial Abilities with Geometric Activities," Young describes a series of graded activities designed to enrich children's spatial perceptions. In "Geometry Links the Two Spheres," Jensen and Spector describe experiences involving whole-body movements that encourage children to think spatially.

Improving Spatial Abilities with Geometric Activities

By **Jerry L. Young**

The role of geometry in the elementary school curriculum is a topic of continuing discussion. Even though every elementary mathematics text includes some geometry, except for measurement, there seems to be little consensus as to which geometric topics should be included and how they should be organized.

One of the roles for geometric activities is to enhance the student's spatial abilities. Spatial abilities encompass many aspects of interpreting our environment, such as, interpreting and making drawings, forming mental images, and visualizing movement or changes in those images.

Why are spatial abilities important? For one thing, our knowledge of the world is influenced by our perceptions, how we interpret and organize visual stimuli. Secondly, many lessons in arithmetic (as well as other subjects) utilize pictures and physical materials that must be interpreted and used in specific ways. A study by Guay and McDaniel reported a positive relationship between the level of achievement in mathematics and spatial abilities with students of elementary school age. One of the hypotheses about the factors associated with lower mathematics achievement of females as compared to males involves the superior spatial skills of the American males. (See *Women and Mathematics*, pp. 90–97.)

Jerry Young is an associate professor of mathematics at Boise State University in Boise, Idaho, where he teaches mathematics content courses for prospective elementary school teachers and mathematics methods for secondary school teachers.

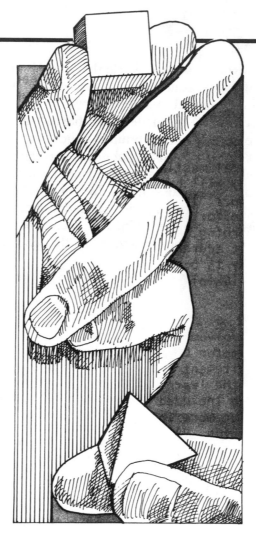

As adults we often forget that the ways we perceive are shaped by our experiences, culture, and education. We do not receive figural input in a neutral manner; we use many learned schema for obtaining, interpreting, and organizing the information we receive. The traditional view has been that our individual spatial abilities are generally static and unchangeable. Recently, more researchers are coming to believe that experience affects these abilities. Soviet mathematics educators have redesigned their ge-

ometry curriculum, in part, for just this reason. For a discussion of changes being made in the geometry curriculum and their underlying psychological basis, see Wirszup's article, "Breakthrough in the Psychology of Learning and Teaching Geometry." One of Piaget's premises is that spatial representations are built upon a series of "actions" on our environment. For Piaget, these "actions" are mental as well as physical. In a cross-cultural study of spatial ability, Mitchelmore (p. 170) hypothesizes that the use of manipulatives in instruction may enhance spatial abilities.

This article includes some geometric activities that teachers can use to give their students experiences that will influence their spatial abilities. The activities are arranged in approximate ascending order from kindergarten through grade eight. Their purpose is to improve spatial abilities, not increase knowledge, so individual responses reveal only spatial abilities and should not be used to judge student achievement.

Activities one through three involve the association of an object with its picture or drawing.

1. Using a set of objects and a picture or drawing of one of them, have the student select the object pictured. Other students could participate by mentally choosing the object. This activity involves interpreting the picture or drawing of a three-dimensional object.

2. As an adjunct to activity one, put an object where the student can feel the object but not see it. By feeling the object, ask the student to identi-

fy a picture or drawing of the object. At a higher level, have the student describe the features of the object and let another student translate this description to choose the picture or drawing of the unseen object.

3. Using a picture of the object (such as a cube) viewed from a variety of positions, have the student hold the object so that he or she is viewing the object from the position shown in the picture. (See fig. 1)

Activities four and five use puzzles to develop abilities to recognize and mentally recall the shape of objects.

4. Use a puzzle with a piece missing and have students decide which piece of a set of pieces will fit into the hole. Let the students check their guess by trying to fit the piece into the hole.

5. Tangrams: Using the seven pieces given in figure 2, try to make the shape pictured.

When doing the puzzle activities, the students should be encouraged to try to imagine if the piece will fit before it is used. This mental process is an essential part of these activities if the students are to develop spatial ability.

Another important visual skill is the ability to see a shape as part of a larger figure. For example, can the student recognize triangle *ABD* in figure 3? This skill is the objective of activities such as six and seven.

6. For the segment $\overline{AD}$ (fig. 4), ask the students to list all of the line segments whose endpoints are the marked points, *A*, *B*, *C*, or *D*. Many students will not immediately recognize segments such as $\overline{BD}$ (There are six segments: $\overline{AB}$, $\overline{AC}$, $\overline{AD}$, $\overline{BC}$, $\overline{BD}$, and $\overline{CD}$.)

Similarly, have students name and/or draw each of the triangles that they can find in plane drawings (fig. 5).

7. Use puzzle problems similar to the following: How many triangles are there in figure 6?

When a person sees a drawing, the orientation of the figure may influence

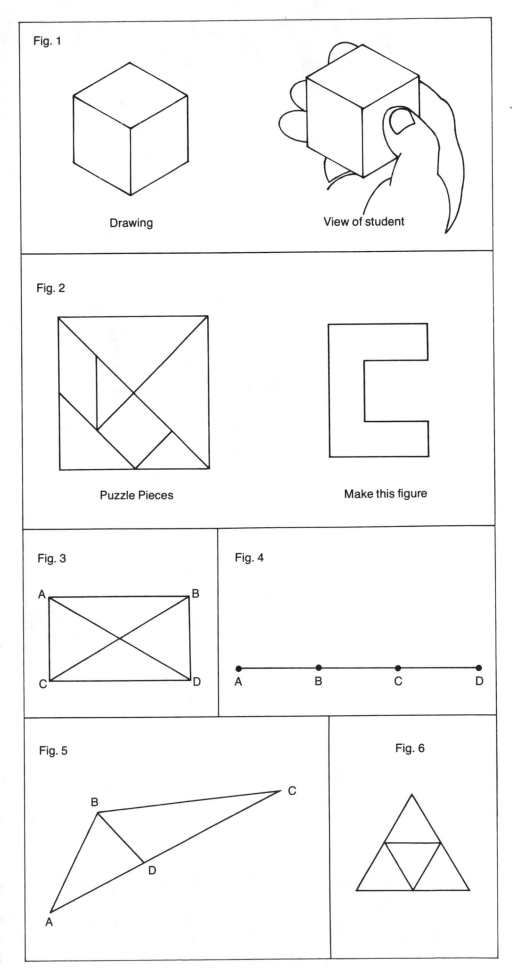

Fig. 1

Drawing   View of student

Fig. 2

Puzzle Pieces   Make this figure

Fig. 3

Fig. 4

Fig. 5

Fig. 6

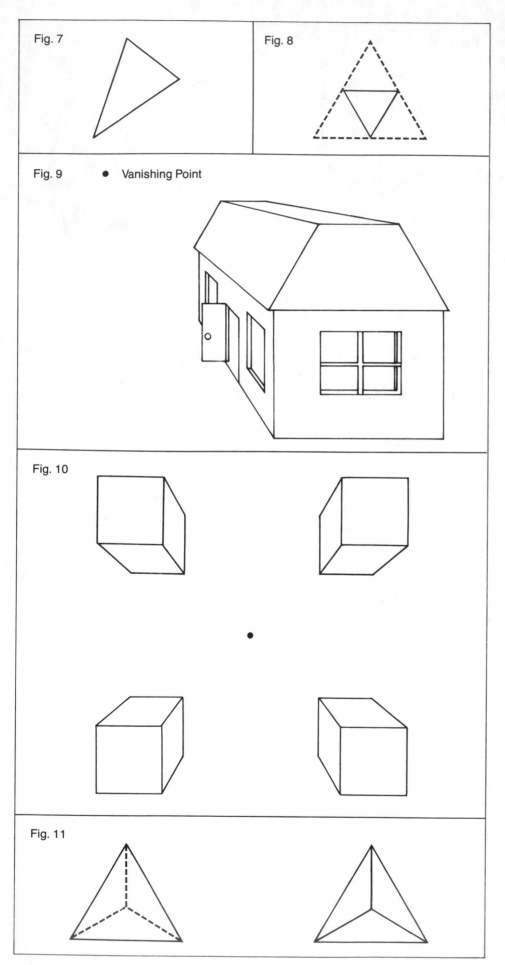

Fig. 7

Fig. 8

Fig. 9 ● Vanishing Point

Fig. 10

Fig. 11

her or his interpretation of that figure. For example, since we usually draw an isosceles triangle with its base horizontal, students are often influenced by this orientation so that they do not recognize the triangle in figure 7 as isosceles. In figure 6, they may not see the triangle highlighted in figure 8. In order to help overcome these fixations on horizontal or vertical orientations, we should vary the orientations of the figures we use as illustrations.

As a culture, we have developed a number of artistic techniques for representing features of an object or scene in our drawings. Our ability to interpret drawings or paintings makes use of the perceptual cues used in these artistic techniques. These cues, such as the representation of depth, are learned through our experiences of drawing and analyzing other drawings and paintings.

The fourth set of activities uses perspective and hidden lines as aids to visualization.

8. Use art lessons to show perspective using a vanishing point (fig. 9). The receding parallel sides of the building all converge at the vanishing point. Students should experiment with the placement of the vanishing point to see how this affects the position of the object relative to the viewer (fig. 10). We should encourage our students to do more drawing and to experiment with different aspects of their art to see what is pleasing and satisfying.

9. In drawing pictures of three-dimensional objects, the relative positions of the edges are often indicated by solid lines for visible edges and dashed lines for hidden edges (fig. 11). Give the students drawings similar to figure 11 and have them hold the appropriate object so it is viewed as drawn. Also have the students draw objects using the hidden edges technique and compare their drawings to determine differences in points of view.

This ability to picture an object from various points of view is the objective of activities ten through twelve.

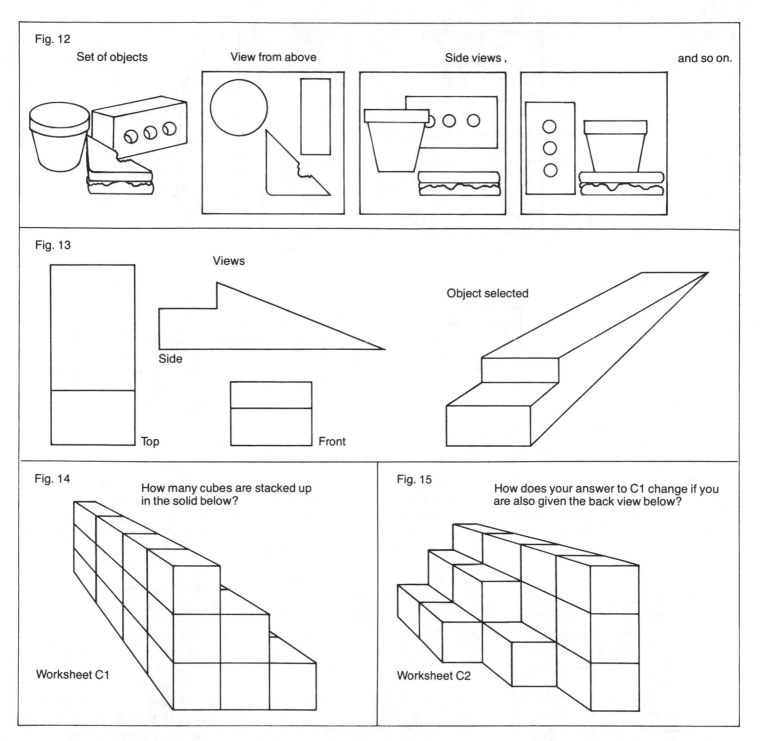

Fig. 12

Set of objects   View from above   Side views ,   and so on.

Fig. 13

Views

Side

Top   Front

Object selected

Fig. 14

How many cubes are stacked up in the solid below?

Worksheet C1

Fig. 15

How does your answer to C1 change if you are also given the back view below?

Worksheet C2

10. There is an activity in the *Environmental Geometry* book from the Nuffield project that gives an overhead view of a group of objects and then asks the student to pick from a set of pictures the view from a particular side (fig. 12). You can make your own set if you do not have access to the Nuffield materials. Have students put the side views around the top view in the position from which that particular picture was taken. They can then check their perception by moving to that position relative to the set of objects.

11. Give the student a sheet of paper on which are shown the top, side, and front views of an object. From a set of objects the student is asked to select the object that corresponds to the given views (fig. 13). At higher grade levels, have the students sketch the different views of selected objects that are simple to draw.

12. Cube counting: Use worksheets similar to those illustrated in figures 14 and 15. When doing the worksheet in figure 14, students have to imagine unseen cubes. The next worksheet gives additional information.

There have been many articles written about investigating symmetries in objects and the teaching of transformations. These can be used with students of elementary school age. The March 1977 issue of the *Arithmetic Teacher* has several such articles. An-

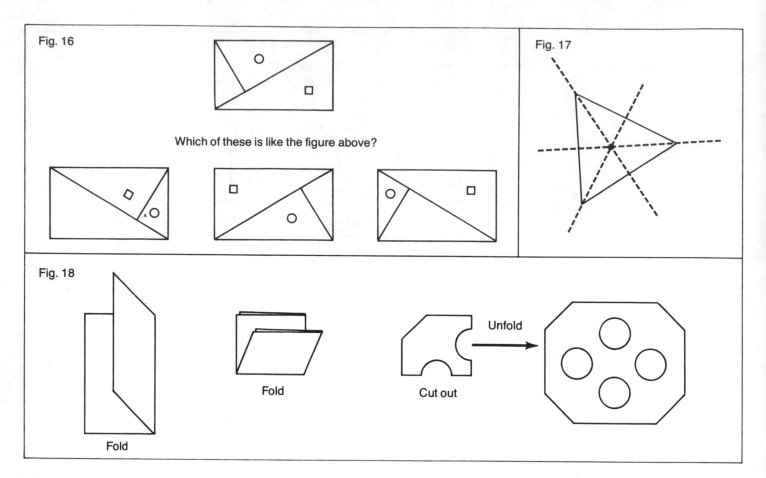

Fig. 16

Which of these is like the figure above?

Fig. 17

Fig. 18

Fold

Fold

Fold

Unfold

Cut out

other of the spatial skills that is very important is the ability to imagine what an object will look like if reflected on a line or rotated about a point. Many tests use items such as the one illustrated in figure 16.

13. Students can collect objects or pictures that have line symmetry or rotational symmetry. The equilateral triangle (fig. 17) has line symmetry on the three different lines since the halves of the figure will exactly match if the triangle is folded (or reflected) on any of dashed lines. Line symmetry can be checked by either using a mirror along the line or folding a drawing along the line.

The triangle in figure 17 also has rotational symmetry about the point at its center. If the triangle is rotated about this point one-third of a turn, the triangle will exactly match its starting position. A rotation of a drawing can be done most easily by using a pin through the point of rotation.

14. Children also enjoy investigating the symmetries of figures formed by folding paper and cutting a design (fig. 18). Have students sketch what they think the result should look like before they unfold it. Give them unfolded patterns and ask them if the patterns could be formed by folding and cutting. Have them verify their answers.

15. Initial activities with transformations can make use of graph paper or geoboards. Figure 19 illustrates the plane transformations of translation (19a), reflection (19b), and rotation (19c) using graph paper. The original position is drawn in red and the final position (image) is in green.

Subsequent transformation activities should involve the students in using a ruler and compass to construct the image. This transition helps students see which properties are necessary when finding an image. For example, the image figure is congruent (same size and shape) to the original figure.

Ultimately the students should be asked to mentally predict where the figure is going to end up and how it will look. They can first sketch the image figure and then check it with a tracing or by construction. Again I emphasize that it is this mental activity that is important in the development of visualization abilities.

16. There are patterns in nature and especially in geometric art that can be analyzed using transformation. The wallpaper pattern in figure 20 can be obtained by reflecting the elementary figure in the dotted lines and translating the result vertically and horizontally. Your students could collect examples of wallpaper, cloth, art works, and so on, that have patterns generated by transformation. Have them analyze these examples for the use of specific plane transformations.

Activities seventeen and eighteen also involve mental transformation, but are not planar.

17. Hexaminoes: Can the arrangement of six squares in figure 21, if cut out and folded along the connected segments, be folded into a cube? Initially, have the students draw the figure on large-square

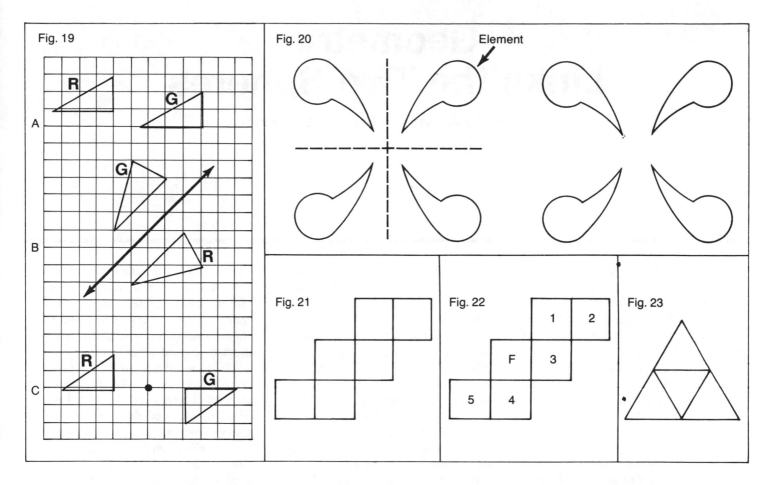

Fig. 19

Fig. 20   Element

Fig. 21

Fig. 22

Fig. 23

graph paper, cut it out, and fold it to see if they get a cube. Later encourage them to mentally fold the figure, cutting and folding as a check. How many different such arrangements of six squares, connected along an edge, will fold into a cube? The term "different" in the foregoing question means that the arrangement cannot be reflected or rotated to match.

Labeling some of the prospective faces in activity seventeen may help visualize the effect of folding. In the example in figure 22, square *F* represents the front of the cube (or face toward the viewer). Square 1 will thus become the top of the resulting cube; square 2, the back; square 3, the right side; square 4, the bottom, and square 5, the left side. This configuration *does* fold into a cube.

18. Have your students make plane patterns that they can use to fold into various solids. For example, figure 23 shows a possible pattern for a tetrahedron. Have students use their patterns to make models of a number of different solids.

The activities that have been described only partially illustrate the experiences that can be provided for students in grades kindergarten through eight. These activities can be expanded upon and other activities can be added to them. The aspects of these experiences that do most to enhance visual-spatial abilities are the interpretation of pictoral information, mental transformation of the information, and the organization of the information according to selected organizational patterns (such as symmetry).

To obtain further information and ideas on the importance of mental imagery, see the bibliography at the end of this article.

References

Arithmetic Teacher 24 (March 1977)

Guay, Roland B. and Ernest D. McDaniel. "The Relationship between Mathematics Achievement and Spatial Abilities among Elementary School Children." *Journal for Research in Mathematics Education* 8 (May 1977):211–15.

McKim, Robert. *Experiences in Visual Thinking*. Monterey, Calif.: Brooks/Cole Publishing Co., 1972.

Mitchelmore, M. "Cross-Cultural Research on Concepts of Space and Geometry." *Space and Geometry*. Columbus, Ohio: ERIC Center for Science, Mathematics, and Environmental Education, 1976.

Nuffield Mathematics Project. *Environmental Geometry*. New York: John Wiley and Sons, Inc., 1969.

Wirszup, Izaak. "Breakthroughs in the Psychology of Learning and Teaching Geometry." *Space and Geometry*. Columbus, Ohio: ERIC Center for Science, Mathematics, and Environmental Education, 1976.

Women and Mathematics, Research Perspectives for Change. NIE Papers in Education and Work, #8, 1977.

AUTHOR'S NOTE: Additional resources

Brydegaard, Marguerite and James E. Inskeep, Jr. *Readings in Geometry from the Arithmetic Teacher*. Reston, Va.: National Council of Teachers of Mathematics, 1970.

Dana, Marcia E. and Mary Montgomery Lindquist. "Let's Try Triangles." *Arithmetic Teacher*, 26 (September 1978).

Mathematics Resource Project. *Geometry and Visualization*. Palo Alto: Creative Publications, 1977.

Sanok, Gloria. "Living in a World of Transformations." *Arithmetic Teacher* 25 (April 1978).

Walter, Marion I. *Boxes, Squares and Other Things: A Teacher's Guide for a Unit in Informal Geometry*. Reston, Va.: National Council of Teachers of Mathematics, 1970.

Geometry
Links the Two Spheres

By **Rosalie Jensen** and **Deborah C. Spector**

When the educators of today were growing up, little information was available about the brain and its relationship to the acquisition of knowledge. Recent research has shown that the brain functions as two hemispheres, a right side and a left side. Each side influences our learning and emotions and provides the foundation for complementary but different types of learning.

The research of Nobel laureate Roger Sperry on the brain offers some insights into the functions controlled by the two hemispheres. Sperry found that when the cablelike structure (*corpus callosum*) connecting the two hemispheres is cut, then communication between the two sides ceases, although each continues to communicate in its particular mode. "The left side of the brain no longer knows what the right side is doing, yet the speaking half of the patient, controlled by the left hemisphere, still insists on finding excuses for whatever the mute half has done, and still operates under the illusion that they are one person" (Pines 1973).

In essence, Sperry and subsequent researchers found that the two hemispheres specialize in different types of cognitive functions (Galin 1976). The left side specializes in logical cognition and uses words as its preferred tool; the right side is holistic and is particularly suited for spatial relationships and music.

Although the particulars on how the two cerebral hemispheres work together are not known, a growing body of research shows that they can enrich each other. The implication is that a student's understanding of mathematics can be affected in a positive manner by developing the creative spatial skills of the right hemisphere.

At least one mathematical skill, the ability to analyze spatial relationships, is a right- rather than left-brain function. According to M. S. Gazzaniga, manipulative skills and geometrical competence occur before left-hemisphere linguistic competence (Sagan 1977, 172). Thus geometry provides a good starting point for the use of movement in mathematics.

Moving into Geometry

Movement is the key to spatial awareness. From infancy on we move out in ever-widening circles, exploring all that is within our grasp.

Each of us lives in both a general space and a personal space. General space is shared with other people, such as in classrooms and shopping centers. Personal space is occupied individually but is included within general space.

The movement activities for students that follow should begin by having each child define his or her personal space. Identify obstacles that must be moved so that each child has the freedom of movement necessary to perform the exercises. Impress children with the fact that for the activities to be successful, each individual must respect the personal space of the others. Therefore they should maintain ample space between themselves and others.

Children have a natural affinity for creative expression. Today's children are particularly excited about the idea of occupying and flying personalized space capsules such as the ones they see on television and in video games. These images can motivate children to explore three-dimensional figures and can reinforce instruction about geometric properties.

When working with younger children, teachers should either have representations of solids available or provide such objects as boxes (cubes), ice-cream cones, pencils, beach balls, cans, and so on. The following sample instructions lead children to explore personal space in the form of a cylinder. The instructions can be changed to meet the particular needs of students. Large groups of children can be divided into smaller subgroups for movement activities while other children are engaged in manipulative activities related to the lessons on solids. All the activities should be adapted to the developmental levels of the children.

Inside a cylindrical space capsule

The movements should emphasize the properties of a cylinder, such as its two parallel surfaces ("top" and "bottom") and its circular cross section. The directions are written to be spoken to a group of children as soft music plays in the background. The

Rosalie Jensen is a professor and chairperson of the curriculum and instruction department at Georgia State University, Atlanta, GA 30303. She teaches undergraduate and graduate courses in early childhood, elementary, and middle school mathematics. Deborah Spector is the public information officer for the National Parents Resource Institute for Drug Education at Georgia State University.

Fig. 1 Hold your arms parallel to the ground.

instructor faces the group and demonstrates each movement.

• Hold out your arms at each side at shoulder height like this. We can say that your arms are *parallel* to the floor and to the ceiling (fig. 1).

• Move your arms around slowly, keeping them parallel to the floor, so that your fingertips make big *arcs* around you. Stretch your arms as you move slowly.

• Imagine that you are in something like this round oatmeal box. (Point to a box.) This is your personal *cylinder*. Pretend that your fingertips are just touching the inside of your cylinder. Now close your eyes and enjoy touching your cylinder.

• Now touch the inside of the top of your cylinder. Put both arms straight up with your fingers pointing up. Your arms are parallel to each other. Stretch one arm as high as you can, very slowly, and then stretch the other.

• Place your arms down at your sides so that they are again parallel to each other. Relax your shoulders and arms.

• Inside your personal cylinder, draw the biggest circles possible by rotating your arms slowly. Go to the front, then up in the air to touch the top, then way back, and down. Now make big circles slowly in the opposite direction—back, up, front, and down.

• Slowly make circles with your head by rotating it to the front, right, back, and left. Make circles slowly in the opposite direction.

• Draw circles with other parts of your body: knees, ankles, upper body, hips.

• Draw big circles slowly above your head with one hand. Draw big circles slowly with both hands anywhere within your cylinder.

• Draw a big circle on the floor with the toes of one foot. Trace the circle several times. Draw big circles of the same size with the toes of the other foot.

After the teacher-directed activities, the children should be allowed to create movements of their own within the cylinder. Remind them that they can make movements in various positions (standing, sitting, kneeling, etc.) and at different levels (high, medium, low); however, they should sense the boundaries of the cylinder and design their movements to stay within these imaginary "walls."

Next design a series of adventures in which children pretend that they are suspended inside space capsules. What kind of movements can be performed within each figure? What movements are performed in touching the inside surfaces of each figure? The edges (when appropriate)? The vertices? Although children may have difficulty drawing space figures and describing their properties, they *are* able to communicate through body movements an intuitive understanding of such concepts as *straight*, *parallel*, *perpendicular*, *angle*, *vertex*, *edge*, and so on.

Inside a spherical space capsule

• Pretend that you are holding a beach ball. Move your hands around to show the shape and size. A ball is a model of a *sphere*. Put it down. Does it roll? Place it somewhere so that it does not roll. Pretend that you have a small ball in one hand. Curl your fingers around it to show its shape and size (fig. 2).

• Stand up and locate a personal space. Pretend that you are suspended in a space sphere that is large enough to allow you to move your arms freely. Only air surrounds you in the sphere. You feel very light, and your movements are smooth and light. You are playing in your space capsule. This one is a sphere!

• Reach forward with one hand until you touch the surface in front of you. In a continuous movement bring your arm overhead and twist your body as you make an arc. Feel the inside of your sphere with your fingertips. (Repeat this movement and subsequent exercises several times with each arm or each leg.)

• Bend forward and move your fingertips down along the inside surface of the sphere and come up along the surface behind you. Continue overhead and down in front of you until you return to your starting point. Did your fingertips trace a circle?

• Balance on one leg as you slowly raise the other leg, touching the inside of your sphere with your toes. (Use a chair or the wall to balance if necessary.) Balance on one leg as you swing the other leg slowly backward and forward to trace arcs inside your sphere.

• Invent movements that you can do inside your sphere. Is your capsule large? How large? Can you make it larger? Smaller? Oops, don't make it so small that you cannot fit inside! (See fig. 3.)

Inside a cubical space capsule

• Pretend that you have a box in front of you that is like a cube. (Provide large boxes for young children.) Feel the sides. Are the sides straight? Are any sides *parallel* to each other? We call the sides of a cube its *faces*. Show me other faces by moving your hands. How many faces are there? (In a similar manner introduce the concepts of *edge* and *vertex*.)

• Find a space facing a wall. Stand with your back straight and your fingertips touching the wall at shoulder height. Move your fingertips up the wall without moving your feet forward. Which parts of your body must stretch? Move your fingertips down the wall to the floor without moving your feet. Which parts of your body bend? Stretch?

• Pretend that you are inside a space cube! Put your fingertips on the surface in front of you. Is it like a

Fig. 2 You are holding a beach ball.

Fig. 3 Where is your sphere? Can you fit into it?

GEOMETRY FOR GRADES K–6

wall? Move your fingertips up and down the surface slowly. Do not move your feet. Stretch and bend as you did when you touched the wall.

• Touch the surface in front of you. Move your hands up. Step forward and move your fingers as far along the upper inside edge as possible. Move your fingers back to the middle and step back as you move your fingers down the face to the lower edge. Move your fingers along the lower edge and back to the middle. Return to your starting position. (Explore the other lateral faces, the top face, and the lower face through similar movements.)

• How many faces does your space cube have? As I count to six, touch a different face on each count. This time as I count to six, touch the faces in a different order.

• How many vertices (corners) does your cube have? As I count to four, jump up and touch a different vertex on each count. Now as I count from five to eight, leap to touch a different vertex with one foot on each count.

• As the music plays, move inside your cube in any way you want. When the music stops, hold your position. Check to see that you are still inside your cube or else you won't be able to get back to earth!

Conclusion

Abstract geometric ideas developed through movement and imagination can unleash cooperative efforts between the two hemispheres of the brain. We hope that your search for ways of teaching geometry through movement will lead you to find other topics in the curriculum that can be enhanced through the use of both cerebral hemispheres.

References

Galin, D. "Hemispheric Specialization: Implications for Psychiatry." In *Biological Foundations of Psychiatry,* edited by R. G. Greneld and S. Gabay, pp. 145–70. New York: Raven Press, 1976.

Pines, M. "We Are Left Brained or Right Brained." *New York Times Magazine,* 9 September 1973, pp. 32–33.

Sagan, Carl. *The Dragons of Eden*. New York: Ballantine Books, 1977. ◖